AI DEMYSTIFIED FOR SENIORS

AN EASY BEGINNER'S GUIDE TO LEARN HOW AI AND SMART DEVICES CAN IMPROVE YOUR DAILY LIFE TO KEEP YOU CONNECTED WITH THE DIGITAL WORLD

GWEN BLAKE

CONTENTS

INTRODUCTION

If you've ever stared at your smartphone and wondered, "Is this thing getting smarter, or am I just getting slower?"—you're not alone. Maybe your grandkids talk about AI like it's an old friend, while you wonder if you missed the memo. Perhaps you've noticed your computer suggesting words before you finish typing, or your phone arranging photos by who's in them, and you think, "How does it know?" These small puzzles can make technology feel more like a maze than a tool.

I understand. Technology changes rapidly, and it often doesn't come with clear instructions. Sometimes it even seems like it's moving on without us. But here's the good news: you don't have to be a computer whiz to keep up. And you're not behind—you're right where you need to be to discover something new.

I'm writing this book because everyone should have a chance to enjoy what technology can offer, no matter when they started using it. My passion is helping seniors like you feel more confident and independent with new technology. Over the years, I've spent time with people who felt left out by the digital wave. I've seen the frustration, but I've also seen the excitement when things start to click. That's what keeps me motivated: the "aha" moments when someone realizes technology can be friendly—and even fun.

So, what's this book going to do for you? Simply put, it will take the mystery out of AI and smart devices. It will show you how these tools can make everyday life easier, safer, and maybe even a little more interesting. We'll talk about what AI really is (and what it isn't), how it's already part of your life, and how you can use it to stay connected, save time, and take better care of your health.

Let's clear the air about AI. You might have heard that it's something only young people understand, or that it's too complicated, or even that it's a little scary. You might worry that using AI means giving up control, or that it's all about robots taking over. Here's the truth: AI is just a tool. It's created to help, not replace, people. Most of the time, it's doing simple things—like suggesting a quicker route on your GPS, sorting out spam emails, or reminding you of appointments. You're still in charge.

Learning about AI isn't about becoming an expert overnight. It's about picking up a few useful skills that fit your needs and interests. Did you know you can use AI-powered apps to organize your photos, keep track of your health, or even chat with a virtual assistant who can remind you to take your medicine? These aren't science fiction. They're real tools you can use today, and they're easier to start with than you might think.

The benefits are real. With just a little know-how, you can use AI to stay in touch with family and friends, manage your schedule, find information quickly, and enjoy hobbies in new ways. You might even find opportunities for remote work or volunteering that use your skills and experience, all aided by AI. It's about turning technology into something that supports your goals, rather than something that gets in the way.

Here's how the book works: Each chapter covers a different aspect of AI and smart devices, from the basics to more advanced uses. You'll get clear explanations, real-life examples, and step-by-step guides. At the end of every chapter, you'll find a takeaway—a practical skill or tip you can use right away. The book is designed to be easy to follow, with pictures and checklists.

I know that starting something new can feel overwhelming. That's why I've made sure the book is as accessible as possible. You don't need any special background or advanced technical skills to get started. If you can use a phone or send an email, you're ready. And if you get stuck, don't worry. The chapters are short and focused, so you can come back and review them as often as you like.

This isn't just a book to read—it's a book to use. I invite you to try out the steps and tips as you go. Jot down notes. Ask questions. Share what you learn with friends or family. If something doesn't make sense the first time, that's normal. Keep going, and you'll be surprised by how much you can do.

You are not alone. I know that learning new things takes time, and everyone moves at their own pace. I'll be with you through each chapter, cheering you on and offering support. If you ever feel stuck, remember that every skill you pick up is a victory. And with technology, there's always something new to learn—no one knows it all.

So here's my challenge to you: Be curious. Give yourself permission to try, to make mistakes, and to celebrate your progress. Whether you want to use AI to make life easier, to connect with loved ones, or just to satisfy your curiosity, you're in the right place. The digital world isn't just for the young—it's for anyone willing to learn.

Let's get started together. Your next "aha" moment might be just a few pages away.

CHAPTER 1
BREAKING DOWN AI— WHAT IT IS AND WHY IT MATTERS TO YOU

TECHNOLOGY OFTEN WEAVES itself into our lives without much notice. You might remember when changing the TV meant turning a dial, or directions involved a paper map. Now, you may casually ask your phone for the weather or talk to Alexa about tomorrow's forecast. Moments like your phone suggesting reminders before you think of them may feel uncanny. Still, they're simply examples of AI quietly helping in the background. No robots or flashing lights – just small bits of assistance. If you feel uncertain or "late to the party," you're not alone. Many people share your questions, and this book aims to guide you—no technical jargon or pressure.

"AI" IN PLAIN ENGLISH—WHAT DOES IT REALLY MEAN?

Let's clear away the mystery. AI, or artificial intelligence, mainly describes software that acts as a tireless helper. Imagine a personal assistant that learns your habits and makes life smoother by picking up on your preferences. If you've used a calculator to balance your checkbook, you've experienced a machine taking over mundane tasks. AI builds on this, handling more complex chores by recognizing patterns and preferences.

The central purpose of AI is to save time, reduce errors, and eliminate tedious work. It can handle repetitive jobs—sorting emails, reminding you of appointments, or highlighting photos of your smiling grandkids. Despite worries that AI will make life more confusing, it's generally designed to cut clutter, streamline your tasks, and leave you with a calmer mind.

Some everyday examples include Voice assistants like Alexa or Siri, which can deliver news, set reminders, or check the weather, just by speaking. Photo apps like Google Photos automatically group pictures by faces or events, so you can easily find your family reunion shots. Email programs filter out spam, so most scam emails never even reach your inbox.

While AI often conjures images of Hollywood robots, the reality is much more practical. AI typically lives inside your phone, smart speaker, or favorite app. It helps you locate items using a tracker or plays your favorite song when asked. The drama stays on film; at home, AI is calm and unobtrusive.

Crucially, using AI doesn't mean you lose control. You're always in charge—you give instructions and make decisions. AI recognizes patterns in your behavior and suggests ways to help.

Here are three specific ways AI may appear in daily life:

1. **Voice Assistants**: Say "Alexa, what's on my calendar?" and she reads you your day's plans.
2. **Photo Organization**: Apps group all your holiday photos together automatically.
3. **Spam Filtering**: Most junk email is filtered out before you see it.

Many people don't realize these features are powered by AI. Research suggests that while only about one in three people think they use AI often, over three-quarters actually do—usually without knowing it.

You might wonder if you really need all this or if it's worth learning now. Here's why it matters: AI can help you stay independent by sending medication reminders or even prompting you to turn off the stove via smart devices. It keeps you connected to loved ones through smart video calls and voice-to-text messages and saves you time by handling chores like sorting emails or tracking appointments. Health apps use AI to monitor your medications, sending timely nudges for refills—especially helpful if you manage multiple prescriptions.

It's normal to feel nervous, worried about pressing the wrong button, or about privacy. The best approach is to start small. Try safe apps, use your phone for simple tasks, and check privacy settings if you have concerns. If you're wary about data collection, you can review app permissions or turn off voice recording features—a little care goes a long way (see Council on Aging, n.d., APA list #6; Myla Training, n.d., APA list #15).

Reflection Prompt: Have You Used AI Today?

Think back over your day. Did you speak to a voice assistant, get a phone reminder, or use an app for directions? List two ways technology helped you—chances are, AI played a role.

You don't need a technical background to use these tools. Every new skill brings more independence and confidence. This book will walk you through the process, one practical example at a time.

HOW AI IS DIFFERENT FROM REGULAR COMPUTER PROGRAMS

When you sit down at your computer or pick up your phone, you're probably used to programs that follow the same steps every time. Think of a classic calculator—punch in 2 plus 2, and it always spits out 4. You could ask it a thousand times, and it would never surprise you, never change its response, and certainly never remember that you prefer to work in dozens rather than tens. That's the heart of a traditional computer program. It operates according to a fixed set of instructions written by a human, and it doesn't adapt or learn, regardless of how much you use it. The program is like a recipe card passed down from your grandmother. The steps never change: measure, mix, bake, enjoy. If you want less salt or an extra pinch of cinnamon, the card doesn't adjust itself—you have to make the change.

Now, compare this to artificial intelligence. AI behaves more like a personal chef who pays attention to your reactions. Maybe one day you wrinkle your nose at too much salt, and the chef quietly notes this for next time. Over several meals, the chef figures out just how you like your eggs or how crispy you want your toast. AI works in a similar way, picking up on your choices and making subtle adjustments to serve you better in the future. Instead of sticking to one rigid routine, it adapts, observes patterns, and learns from your actions. The more you interact with it, the more it can customize its help.

For example, when you use Netflix or another streaming service, the AI inside pays attention to what you watch. If mysteries keep popping up on your screen because you just finished a detective show, that's not an accident. The AI noticed your interest and tries to offer more of what you might enjoy next. This isn't magic, and it doesn't mean the program is "thinking" like a person with feelings or desires. It simply tracks patterns in your habits and adjusts its suggestions based on the information it collects.

Think about Google Maps. At first, it offers the typical route from your home to the grocery store. But if you take a shortcut through Elm Street every Thursday, the AI starts suggesting this faster way automatically. It's not because the app "knows" you in the way a friend does. Instead, it picks up repeating behaviors and acts accordingly. Over time, it adapts to make your experience smoother.

A traditional program can't do this. It's bound by its original design—unchanged by your actions, never learning new tricks unless a programmer updates it. In contrast, AI programs have the capacity to "learn" through repetition and exposure to your activities. This learning isn't conscious; it's all about spotting patterns and making predictions.

Take spellcheck on your word processor as another example. In the early days, spellcheck simply compared words to a dictionary—if it wasn't there, it underlined it in red, no exceptions. Modern spellcheckers powered by AI now look at how people actually write— common typos, frequent corrections—and can even suggest words based on what fits your sentence best. If you consistently type "teh" instead of "the," the AI often learns to correct that instantly for you over time.

An AI system doesn't "think" like humans do; it doesn't wonder why you like certain foods or shows. Instead, it processes enormous amounts of data—your clicks, searches, or even pauses while reading —and uses statistical analysis to predict what might help you next. This prediction feels personal because it's based on your previous choices.

It's normal to wonder if AI will overstep or get too personal. However, most systems are designed with privacy in mind and only adjust based on what you allow or share with them. For instance, while Netflix makes suggestions based on shows you've watched on your account, it won't pull preferences from somewhere else unless you connect that information yourself.

A good analogy is this: using traditional software is like following a path set in concrete—no matter how many times you walk it, the scenery never changes. Using AI is more like strolling on a garden path where flowers bloom in response to where you stop and smile. Each walk feels a little more tailored to your interests and preferences.

You don't need to be an expert or have technical skills to see this difference in action. Every time a device seems to "get" what you want without being told explicitly—like suggesting a birthday greeting when your calendar shows a grandchild's birthday—that's AI at work. It's not reading your mind; it's just learning from what you've done before and quietly trying to help make things easier.

AI adapts in small ways that add up over time. The more comfortable you get with these changes and let yourself experiment—by clicking around or trying new features—the more helpful these systems can become for your daily life.

BUSTING THE MYTH—YOU DON'T NEED TO BE A TECH EXPERT

You might hear it all the time: "This newfangled technology is for the young, not for me." Or maybe you've caught yourself thinking, "I'm just not a tech person." That's a myth that keeps too many smart, capable people from enjoying new tools that could make life easier. The truth? AI isn't reserved for whiz kids or folks who grew up with computers. It's been created for everyone, especially for people who want things to be simple and direct. I've seen it firsthand—real people, with little tech background, making AI work for them in ways that are practical and even a little bit delightful.

Take, for example, a grandmother who wanted help remembering her daily medication and keeping track of when to water her houseplants. She didn't need to study computer science or read a thick manual. She set up Alexa in her living room with the help of a family member, then discovered she could say, "Alexa, remind me to take my pills at 8 a.m.," and the device took care of the rest. Her confidence grew each day as she added reminders for birthdays and even grocery lists. She told me, "I just talk, and Alexa listens. It's easier than writing myself sticky notes all over the kitchen." This isn't science fiction; it's everyday life, made easier by a tool that's happy to listen and help.

Another story comes from a retiree who had boxes of old photographs and hundreds more on his phone. Sorting them seemed impossible until he tried Google Photos. With a few taps, he watched as the app grouped pictures by faces and events. Soon, his family albums were organized, and he could pull up every photo of his daughter's wedding or his grandson's first fishing trip in seconds. He said, "I never learned computers in school, but this felt like magic. I just pressed a button, and all my memories fell into place." His experience shows that you don't need technical skills; you need to give yourself permission to try.

These tools are built with simplicity in mind. You're not required to memorize commands or learn complicated steps. Often, you speak or tap once or twice. Take Siri, for example. You don't need to type out a message or fumble through menus—just say aloud, "Send a text to

Mary," and Siri walks you through the rest. The device handles the details while you focus on what you want to say. It feels as natural as asking someone in the room to help you out, only it's your phone doing the work.

MYTH VS. FACT: A QUICK REALITY CHECK

- **Myth:** "I have to understand computers to use AI."
- **Fact:** "You can use AI by simply asking questions out loud or tapping an icon."

It's common to believe that these devices require special knowledge or training. In reality, they're designed to lower barriers, making them accessible to anyone. You don't need to know how they work under the hood—just how to tell them what you want.

Worried about pressing the wrong button? Don't be. Most AI-powered features are forgiving; they offer clear instructions and even let you undo actions if needed. The key is to start with small steps and build confidence from there.

Here's something to try right now if you have a smartphone handy: Wake your device by saying, "Hey Siri, what time is it?" (If you have an iPhone) or "OK Google, what time is it?" (for Android phones). Notice what happens. Did your phone respond with the correct time? If so, you've just used AI—no manuals, no passwords, just your voice and curiosity at play.

Give yourself credit for every success, no matter how small it may seem. Many people your age have found themselves surprised at how natural it becomes after just a few tries. One woman told me she started using her smart speaker just for weather updates, but soon found herself asking for recipes, playing her favorite oldies, and calling her daughter hands-free every Sunday afternoon.

AI's learning curve is often much gentler than expected. If you can speak, listen, or tap a button, you're already equipped to use these features. Allow yourself room for mistakes—they're part of learning—

and remember that most devices come with "help" menus or online support if you want extra reassurance.

Trying new things doesn't mean becoming an expert overnight; it means enjoying small victories along the way. Each tap or spoken command is proof that these tools are for everyone, including you.

EVERYDAY MOMENTS POWERED BY AI (YOU'RE ALREADY USING IT!)

You might be surprised to learn just how much artificial intelligence is quietly woven into your daily routines, often without your knowledge. Think about the last time you opened your photo gallery on your phone and saw that photos of your grandchildren were already grouped in neat little albums. Maybe you've picked up your phone and found it suggesting the name of a family member as soon as you started typing a message, or you've had your thermostat adjust itself when the weather suddenly changed outside. These are not just lucky guesses or clever programming. There are subtle ways that AI steps in to make life less tangled and more enjoyable.

Take Google Photos, for instance. It can scan through thousands of pictures, pick out familiar faces, and create albums for each grandchild. So, instead of scrolling endlessly to find that perfect holiday snapshot from last December, you tap a face, and every birthday smile or backyard adventure is right there. The best part? There's no need for you to label anything or organize files; AI manages the sorting, saving you time and sparing you frustration. Now, consider medication reminder apps. These handy programs monitor your schedule and pop-up friendly nudges, so you never forget a dose. No more sticky notes all over the fridge—now, a soft chime is all it takes to remember.

Home comfort gets an upgrade, too. Smart thermostats learn from your patterns. If you turn down the heat every night at the same hour, the thermostat will eventually do it on its own. When an unexpected cold snap hits, the system adjusts to keep things cozy without you lifting a finger. It's like having a caretaker who never forgets, never complains, and always wants your home just the way you prefer.

Invisible helpers also keep you safe online. Email spam filters are one of those silent guardians. Dangerous or annoying messages rarely bother you because these filters scan every email, spot suspicious ones, and send them straight to junk. It's not luck—it's AI working behind the curtain to catch scams and clutter before you even see them. If you've noticed fewer bizarre messages about lottery wins or urgent bank requests, thank the AI that's hard at work in your inbox (see Clean Email, n.d., APA list #5).

Typing on a smartphone can be awkward, but have you noticed how your device often guesses what word you want next? That's predictive text in action. As soon as you start writing "Happy bi...", your phone suggests "birthday" before you finish. It doesn't read your mind; it remembers common phrases and helps you along, making texting faster and less stressful.

AI hides in other places, too. Maybe you've used a navigation app that rerouted you around traffic or construction without needing to ask for an alternate path. Or perhaps your smart speaker played music based on your listening habits during dinner, picking out tunes that suit your mood. Each of these features relies on AI to notice patterns and offer solutions before you realize you need them.

If any of this sounds familiar, give yourself some credit—you're already using AI more often than you think. Every time a notification pops up with just the information you need, that's AI lending a hand. The fact that these features feel so seamless is proof they're working as intended: quietly in the background, making life smoother.

Sometimes it helps to see things spelled out plainly. Here's a simple checklist to build confidence and spark curiosity for what comes next:

Checklist: Are You Already Using AI?

- Have you seen photos grouped by faces or places without doing it yourself?
- Does your phone or tablet send reminders for medications or events?

- Have you noticed fewer scam or junk emails in your inbox lately?
- Does a device in your home automatically adjust the lights or temperature?
- Did your phone suggest words while texting or emailing?
- Have you received a suggested alternate route while using maps?
- Did you ever use voice search to find something on Google?
- Have you played music by just asking a smart speaker?

If you checked off even one box, congratulations—you're already taking advantage of AI-powered tools in everyday life! Each small victory means you're more capable than you might have realized, and there's plenty more to explore ahead.

WHY LEARN AI NOW? STAYING INDEPENDENT, SAFE, AND CONNECTED

Reaching retirement or simply having more time on your hands often opens the door to new possibilities. Some people travel, others dive into hobbies, but there's a growing number who want to keep their minds sharp and maintain strong connections with loved ones. This is where learning about AI fits perfectly. It's not just about gadgets and buzzwords; it's about preserving your independence, protecting your well-being, and making life more enjoyable. Imagine being able to manage daily health routines or household tasks without constant worry. Medication reminders pop up on your phone, so you never skip a dose. Smart lights and thermostats adjust the environment for your comfort and safety—no more fumbling in a dark hallway at night. These features aren't far-fetched or reserved for the tech-savvy. They're practical, easy to use, and can give you far greater control over your routine.

Staying in touch with grandchildren, siblings, or old friends takes on new meaning when you tap into smart communication tools. Video calls used to be complicated, but now you can launch them with a few spoken words or a gentle tap. Some platforms even help translate messages between languages, making it easier to connect if your family is scattered across the globe. You're not limited by distance or complicated setups; technology brings people together in ways that feel natural and warm. A group chat with photos and updates from family members becomes a breeze, while voice messages let you share stories in your own words—no typing needed. These small advances add up, helping you feel part of the action and not left on the sidelines.

Managing health becomes much simpler with the support of AI-driven apps. Keeping track of doctor appointments, prescription schedules, or daily exercise can feel overwhelming when handled on paper calendars or sticky notes. A single app can organize all those details, issuing reminders and even letting you send questions to your provider through secure portals. Many seniors find peace of mind knowing they have a digital assistant watching out for missed pills or doctor visits (see GoodRx Health, n.d., APA list #11). Technology also acts as a powerful ally against scams. Email and messaging apps use AI to scan for suspicious links or fake offers, giving you a heads-up if something looks risky (see Clean Email, n.d., APA list #5). This silent protection means you can focus on genuine messages from friends without second-guessing every email.

Saving time isn't just about convenience—it's about giving yourself space for what matters most. Instead of sorting through dozens of emails, AI helps flag what's important. Smart calendars rearrange appointments when conflicts pop up, so you don't have to play phone tag. For many, this newfound efficiency means more hours for reading, gardening, or simply relaxing with a favorite show.

Learning about AI brings another benefit—mental stimulation. Exploring a new topic keeps your brain agile and introduces fresh challenges to solve. Taking part in online classes or virtual book clubs through video platforms powered by AI isn't just about gaining knowledge; it's also about joining conversations and building community. Brain games serve as a fun way to challenge yourself and even compete with friends in friendly matches (see Galleria Woods Senior Living, n.d., APA list #8). These activities nurture curiosity and strengthen memory, creating daily opportunities for growth.

Engaging with the community is easier than ever before. Many senior centers now offer virtual meetings, hobby groups, or discussion circles hosted through smart devices. You might join an online painting class or participate in a neighborhood watch meeting from your living room, all thanks to accessible technology. These opportunities foster relationships with new people while nurturing existing bonds—a win for both social well-being and mental health.

"It's too late for me to learn something like this," is a phrase I hear often, but it rarely holds true for long. I remember speaking with a retiree who hadn't touched a computer until her late sixties. She started learning how to use a smart speaker after her grandchildren gave her one for Christmas. At first, she only asked for the news or weather. Soon enough, she was setting reminders for family birthdays and playing audiobooks aloud as she knitted by the window. The confidence she gained from those first steps rippled out into other parts of her life—she tried online classes for watercolor painting and reconnected with an old friend through video chat. That spark of curiosity transformed quiet afternoons into lively ones.

Making the leap into new technology isn't about chasing fads or feeling pressured by others; it's about choosing tools that make your days brighter and more manageable. The world doesn't stand still— and neither do you. With every small success, whether it's sending your first video message or joining a digital book group, you're building confidence and proving that learning never has an expiration date.

COMMON FEARS ABOUT AI—AND HOW TO OVERCOME THEM

If you've ever caught yourself thinking, "I'm just not good with gadgets," or felt a wave of anxiety as you hovered a finger over a new button, you're in good company. Many older adults share concerns about technology, and artificial intelligence can sound especially intimidating. I hear worries all the time: "What if I press the wrong button and mess everything up?" or "It feels like these devices are always listening—are they spying on me?" Even more common is the fear of falling victim to a scam or unwittingly sharing private information.

These thoughts are not only understandable but also completely normal, especially when every week seems to bring news of another online scam or data breach. The truth is, you're not alone in these feelings—far from it.

Let's start with the fear of breaking something. There's this idea that one slip could ruin your device or lose precious memories. The reality is, most smartphones and smart speakers are built to be forgiving. It's nearly impossible to "break" your device with a single tap or voice command. If you do something you didn't mean to, there's usually an easy way to undo it. Think of it like learning a new appliance in the kitchen; mistakes might make a mess, but rarely cause lasting harm. Practicing with harmless features is a gentle way to build confidence. For instance, try setting a simple reminder—something like, "Remind me to water the plants at ten." If it doesn't work the first time, no harm done; just try again. Each small success chips away at hesitation.

Privacy is another big concern, and you might wonder if using AI means your personal life is suddenly up for grabs. "I don't want strangers knowing where I live or what I say," is a thought I hear often. The good news is that privacy settings put you in control. On your smartphone, you can check which apps have permission to access your location, microphone, or camera. Go into your phone's settings and look for "Permissions" or "App Settings." If you see something that doesn't feel right—maybe an app wants your location but shouldn't need it—simply turn off that access with a tap. Manufacturers have started marking trustworthy apps with special icons or "Safe to Try" badges, so you know they've been reviewed for security (see Council on Aging, n.d., APA list #6). Sticking to well-known brands and downloading apps directly from your device's official store reduces your risk even further.

Scams are another worry. Many people hesitate to try new features because they've heard stories of friends or relatives getting tricked by fake emails or phone calls. Here's the thing: AI can actually help protect you from these dangers. For example, email apps use AI to filter out suspicious messages before you see them. If something does make it through and asks for money or personal details, pause and ask someone you trust before responding. Never give out personal information to anyone who contacts you unexpectedly—whether it's by email, phone call, or text message.

Some people picture AI as an all-seeing eye, always listening and recording every word. That's more fiction than fact. While smart speakers do wait for a "wake word" like "Alexa" or "Hey Siri," they aren't recording everything in your home. You can review or delete voice recordings in the device's settings if you ever feel uneasy. If privacy remains a concern, turn off the microphone when not in use—a physical switch on many devices makes this easy. You're always in the driver's seat; these features exist for your convenience, not to snoop on your life.

A big myth is that AI is so complicated that only experts can use it safely. In reality, AI tools are designed with everyday people in mind, including seniors who want simplicity and reassurance. You don't need to understand how AI works on the inside; you need to know how to use what's useful to you.

Here's something practical: choose one trusted app or device and explore it at your own pace. Start with small actions—a reminder, a weather check, or listening to music by voice command. Celebrate each time it works as planned. Bit by bit, these tiny steps build both skill and trust.

Learning about AI isn't just about keeping up—it's about staying safe and independent as our world changes. The more you know, the more power you have over your digital life. Every attempt builds confidence and increases your ability to spot trouble before it starts. You'll find that many fears fade once you see how much control you truly have.

Remember: This book will keep guiding you with step-by-step help and practical tips as we go along. Curiosity and caution both have their place—and together, they make you stronger than any technology challenge ahead.

CHAPTER 2
MEET YOUR AI HELPERS
—SMART DEVICES IN
YOUR DAILY ROUTINE

GETTING TO KNOW VOICE ASSISTANTS—ALEXA, SIRI, AND GOOGLE ASSISTANT

IMAGINE HAVING a friendly helper at home, ready to answer questions, remind you of tasks, or offer a quick joke when you need it. This isn't science fiction—it's the daily reality for millions, thanks to voice assistants. These tools aren't intimidating robots or complicated gadgets. They're simple, voice-activated helpers designed to listen, understand, and respond to your requests in plain language—just as you'd use with a neighbor or friend. Each assistant has their own style and strengths. The three main options are Amazon's Alexa, Apple's Siri, and Google Assistant.

Alexa is like a cheerful companion, constantly prepared with an answer or recipe. Alexa lives inside Amazon Echo smart speakers— those small, round, or cylindrical devices often found in kitchens or living rooms. She's also available in smart displays for video calls or step-by-step help. Siri, Apple's voice assistant, feels like an old friend for anyone with an iPhone or iPad. Siri is courteous and quick, perfect for reminders or sending messages hands-free, and is already available on Apple products—just say "Hey Siri." Google Assistant, from Google, is quick to find information and answer nearly any question.

You'll find Google Assistant on Android phones, Google Nest Hubs, and various smart speakers and tablets. Each brings something special —Alexa excels at routines and smart home control, Siri is great for Apple apps and privacy, and Google Assistant is a whiz at searching and scheduling tasks.

These assistants are accessible through many devices. Alexa works best with Echo speakers like the Echo Dot and the Echo Show. Siri is built into iPhones, iPads, and Macs. Google Assistant operates on Android phones, Google Nest Hub displays, and some smart TVs and other speakers. You may already have one—or can easily add one to your home.

Voice assistants do more than just answer questions—they help with daily needs that can easily be forgotten. For example, if you want help remembering medications, just say, "Alexa, remind me to take my morning pills at 8 AM," for a daily reminder (GoodRx Health, n.d., APA list #11). Planning to go out and curious about the weather? Ask, "Hey Siri, what's the weather today?" for an instant local forecast. Prefer news with your morning coffee? "OK Google, give me today's news update," and Google Assistant reads headlines aloud while you relax.

Privacy is essential, and you might wonder if your assistant is always listening or recording everything you say. In reality, these assistants only start processing your requests when they hear their wake word —"Alexa," "Hey Siri," or "OK Google." Otherwise, they're idle in the background. You control when the assistant listens. With Amazon Echo, there's a physical microphone mute button—press it, and Alexa stops listening until you unmute (Council on Aging, n.d., APA list #6). On iPhones or iPads, you can turn off "Hey Siri" in settings with a tap. Google devices have an easy-to-access mute icon as well.

It's natural to feel awkward speaking to a device, but these assistants are designed to keep things simple. They're patient, don't judge mistakes, and will repeat responses if needed—the goal is to free you from small chores, making life smoother.

Reflection Prompt: Try It Out!

Think of something you often forget or wish were simpler—maybe remembering when a favorite show airs or calling family on Sundays. If you have Alexa, Siri, or Google Assistant, ask it to remind you or make the call. Jot down how it went: Was it easy? Did the assistant get it right? What other tasks might you try next time?

Voice assistants aren't about taking over—they're about helping where you want, giving you more time and energy for what matters most.

SETTING UP A SMART SPEAKER—A STEP-BY-STEP GUIDE FOR BEGINNERS

Opening the box of a new smart speaker feels a bit like getting a present—there's curiosity, a little nervousness, and maybe even a tangle of cords that makes you pause. Don't worry, you don't need to be a gadget expert. Most smart speakers, like Amazon Echo or Google Nest Mini, keep things straightforward. First, find a good spot for your device: somewhere central is best, so it can hear you from the kitchen, living room, or wherever you spend time. Plug in the speaker with the cord provided. A small light ring or dot should appear—on an Echo, it glows blue at first, then shifts to orange or another color as it starts up. The Google Nest Mini might show a line of white lights. These lights aren't just for show; they're little signals from your new helper, telling you it's waking up and getting ready to meet you.

Once you see those indicator lights, you know the device is alive and listening for instructions. The colors have meaning: blue often means it's thinking or processing; orange or yellow may signal it's trying to connect to Wi-Fi. If you see red, on most devices, that means the microphone is muted—so if you want to talk, tap the mic button again to turn it back on. Hearing a chime or seeing lights swirl usually marks the start of setup mode.

Now comes the part where you connect your speaker to Wi-Fi and your phone or tablet. Grab your smartphone or tablet—you'll need it to finish setup. Search for the Alexa app (for Amazon Echo), Google Home app (for Google Nest), or Apple Home app (for some HomePod

Indicator lights – what they mean

○ Solid ring
○ Listening
↻ Processing
● Notification
🎤 Microphone muted
▯ Blinking bar

speakers) in your device's app store. Download and open the app; don't worry if it takes a minute to install. Once open, the app will guide you through finding your speaker. The screen may display large icons and text—if not, most phones let you zoom in with two fingers for bigger print. You'll see prompts like "Add Device" or "Set Up New Device." Tap these and follow along; the app will find your smart speaker nearby, especially if it's plugged in and showing those setup lights.

Step-by-step:
Download the Alexa, Google Home, or Apple Home app

1. Open App Store (iPhone) / Open Google Play (Android)
2. Type in the search box
3. Select the official app
4. Tap GET / INSTALL, then OPEN

When asked to connect to Wi-Fi, the app will either show a list of available networks or prompt you to type in your Wi-Fi password. Here's a tip: have your password written on a sticky note nearby for easy access. Many apps now offer a large-print option or let you copy and paste pass-words from another note, making things easier on your eyes. If typing is tough, ask a family member or friend to help—there's no shame in teamwork.

Personalizing your smart speaker makes it feel more like yours and less like a stranger in the room. The setup app usually gives you a chance to add your name, set your location (for local weather or traffic), and even pick a profile photo if you like. Entering your address or postal code enables the speaker to provide you with accurate news updates and weather forecasts. Some apps offer the option to choose your preferred news sources or music services—pick what sounds good and skip anything that feels unnecessary. If privacy is on your mind, skip adding a photo and use only your first name.

Sometimes, things don't go as planned—the Wi-Fi won't connect, or the speaker seems stuck on a spinning light. If this happens, don't panic. First, check if your internet is working by testing another device, like your phone or laptop. If other devices are fine but the speaker won't connect, unplug the speaker from the wall, wait ten seconds, then plug it back in. Many small hiccups fix themselves with this simple reboot. If the device freezes and won't respond at all, look

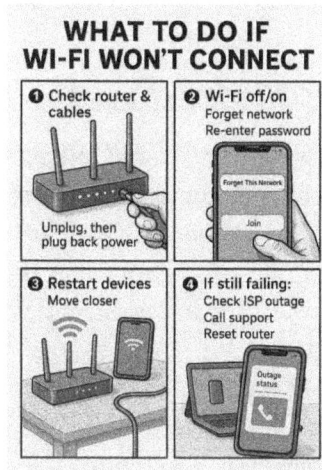

WHAT TO DO IF WI-FI WON'T CONNECT

❶ Check router & cables
Unplug, then plug back power

❷ Wi-Fi off/on
Forget network
Re-enter password

❸ Restart devices
Move closer

❹ If still failing:
Check ISP outage
Call support
Reset router

for a tiny reset button on the bottom—pressing this gently with a paperclip will start fresh (the setup app will walk you through this too). Double-check your Wi-Fi password; even one wrong letter will stop things from connecting.

If you're still stuck after these steps, use the app's help section—often marked with a question mark—and search for common errors. You may also find support phone numbers or chat options right in the app, designed for regular folks who run into bumps during setup.

Visual Exercise: Your Setup Checklist

- Find a central spot for your smart speaker
- Plug it in; watch for indicator lights
- Download and open the Alexa, Google Home, or Apple Home app
- Follow the prompts to add your device
- Enter your Wi-Fi password carefully

- Type in your name and location
- Choose a profile photo (optional)
- Test by asking "What time is it?"
- If stuck: unplug and restart; check internet; try the reset button

Take this process one step at a time. Each step brings you closer to having a helpful digital assistant at your side—no rush necessary.

USING VOICE COMMANDS—PRACTICAL PHRASES FOR DAILY LIFE

Getting the hang of talking to your smart speaker or phone is simpler than it might seem. The first thing you'll want to know is how to "wake up" your device. This means grabbing its attention before you speak, just like calling someone's name across the room before you ask them a question. Each assistant has its wake word. For Amazon's devices, say "Alexa." With Apple gadgets, use "Hey Siri." If you're talking to Google devices, say "OK Google" or sometimes "Hey Google." After you say the wake word, your assistant listens closely and waits for your command. You'll often see a little light or hear a soft sound to let you know it's listening—you don't need to shout, just speak normally.

Once you're comfortable with wake words, you can start using voice commands for all sorts of things. To help you remember, here's a simple "cheat sheet" you can print or write down in big letters. Tape it near your favorite chair or on the fridge for easy reference:

- "Alexa, what's on my calendar today?"
- "Hey Siri, remind me to water my plants at 4 PM."
- "OK Google, call my daughter."
- "Alexa, set a timer for 15 minutes."
- "Hey Google, what time is my doctor's appointment?"
- "Siri, send a text to Sam."
- "Alexa, turn on the living room lights."
- "Hey Siri, play Elvis Presley."
- "OK Google, what's the traffic like to the grocery store?"
- "Alexa, what's the weather tomorrow?"
- "Hey Google, tell me a joke."

Don't feel you must memorize them all at once. Try one or two that fit your routine. For example, if you want to check your daily agenda, just say the correct phrase after the wake word. If you'd like to call family without searching for their number, try a calling command. Setting reminders is especially helpful—whether it's watering plants, feeding a pet, or taking medicine.

Speaking clearly is key to getting good results. Pause for a moment after the wake word so your device knows to listen. If there's background noise—like the TV or radio—you might need to speak a bit louder or move closer to your device. Sometimes your assistant might misunderstand; perhaps it thinks you said "call Mary" instead of "call Larry." If this happens, don't worry. Repeat the command slowly or try saying it in a new way. Instead of "call my daughter," try "call Susan's mobile"—the more specific, the better. If your assistant still doesn't get it right, rephrase using simpler words.

Using voice commands can do more than just save time—they can also bring peace of mind and independence around your home. For example, if your hands are full or you're not near a light switch, saying "Alexa, turn on the lights" (if you have smart bulbs) makes moving around safer. You can set up routines like "Good night," which turns off all the lights and locks the door with one phrase. If you're concerned about safety or want quick access to important information, ask your assistant for emergency contact details—"Hey Siri, who is my emergency contact?"—and keep that information updated in your phone or device.

Some folks find it helpful to have their assistant remind them of things that are easy to forget—closing garage doors at night, locking up before bed, or taking a walk after lunch. The assistant never gets tired of repeating itself and doesn't mind being asked the same thing every day. This gentle support can keep you on top of your schedule and reduce stress.

Here's an interactive idea: jot down two or three tasks that often slip your mind or take up too much time. Next to each one, write out the voice command you'd use to make life easier. For instance, if remembering birthdays is tricky, write: "Alexa, remind me of Lisa's birthday every March 8th." If checking the weather before heading out is helpful, write: "Hey Google, do I need an umbrella today?" This little exercise not only builds confidence but also personalizes your use of technology.

For those seeking greater independence, there are ways to take it to the next level. With compatible devices and a bit of setup help from a friend or family member, you can control thermostats, adjust window blinds, or even unlock doors using only your voice. It might sound futuristic, but these options are available now and growing more common every year.

Don't hesitate to experiment and see what works best for you. Start with simple requests and work up as you get more comfortable. Before long, you'll find that giving voice commands feels as natural as picking up the phone or flipping a light switch—often faster and easier, too. Every successful command is another step toward making technology work for you instead of feeling like an obstacle.

CUSTOMIZING SMART ASSISTANTS FOR REMINDERS AND APPOINTMENTS

The real magic of these smart helpers kicks in once you start making them work for your daily routine. You can teach your assistant to remember what matters to you, from medication schedules to special events. Start by thinking about something you'd like to automate—a pill you need to take every morning, or that weekly coffee chat with a friend. Grab your phone, tablet, or smart speaker and speak up. Say, "Remind me to take my blood pressure pill every day at 8 a.m.." The assistant will ask if you want this reminder once or every day. Pick daily, and just like that, it'll nudge you at the right time—no more sticky notes or second-guessing. For appointments, try, "Schedule a haircut with Linda on Friday at 10 a.m.." The assistant will confirm the details and add them to your list. If you want something to repeat, like a weekly bridge game or Sunday lunch, just specify, "Remind me every Sunday at noon."

Don't stop at the basics—these reminders are highly customizable. Adjust the alert tone if you struggle to hear soft chimes. Open your settings or ask your assistant for "volume up." Some devices let you pick sound types, from a gentle bell to a more energetic beep. If hearing is a challenge, many assistants and smart displays offer visual cues as well. Turn on flashing lights or screen notifications to ensure you don't miss a reminder, even in a noisy room. These options are easy to find under "Accessibility" in your app's settings or by asking, "Show visual notifications for reminders." You can even set reminders to repeat more often, like every two hours for hydration, or less often, such as the first Monday of each month.

Syncing your assistant with a calendar app is a smart way to keep everything in one place. Connect it to Google Calendar or Apple Calendar so every appointment and reminder shows up together. In the assistant's app, look for "Add calendar," then follow the steps to sign in with your email account. Once connected, say things like, "Add doctor's appointment to my calendar for next Wednesday at 3 PM." These entries appear instantly on your phone or tablet. Prefer paper? Many assistants can print a daily agenda for you if you have a connected printer—say, "Print my schedule for tomorrow." This blend of old-school and new-school keeps you organized without letting anything slip by.

Building routines takes things further. Imagine waking up and saying, "Good morning," then hearing the weather forecast, today's news headlines, a gentle nudge to take your vitamins, and maybe even an inspirational quote to start your day with a smile. You can create this kind of chain reaction by setting up a routine in your assistant's app. Tap on "Routines" or ask your device for help setting one up. Choose actions—such as weather report, medication reminder, or morning greeting—and select a time or trigger phrase, like "Start my day." Hit save, and now every morning starts on the right foot. Some folks like to add music or a favorite radio station as part of their routine. Others prefer an evening checklist—"Good night" can turn off lights, lock doors (if you have smart locks), and remind you about tomorrow's appointments.

You're not limited by what's already in your phone; assistants can adapt as your needs change. Maybe you want reminders that change with the seasons—gardening tasks in spring or snow shoveling alerts in winter. Update or add new reminders as needed. If you're sharing space or caregiving for someone else, create reminders for their medicines and appointments too. The assistant can distinguish between voices on some models, so everyone gets the right message.

If you ever feel overwhelmed by beeps and buzzes, it's easy to adjust notification frequency or even turn off some reminders temporarily. Ask your assistant to pause reminders while you nap or mute notifications during family visits. You can also set "quiet hours"—times when alerts are silenced except for emergencies.

For those who want everything at a glance, try asking for a summary: "What are my reminders today?" or "List my appointments this week." The assistant will read them out or display them on a bright screen.

Personalizing these features can make life feel more manageable and less rushed. A few minutes spent setting up tailored reminders means fewer worries about missing doses or forgetting plans. It might even free up some mental space for hobbies or connecting with friends. And when life changes—as it always does—you can tweak reminders and routines without fuss. The assistant never gets tired of helping you stay on track, letting you decide what matters most each day.

SMART TVS AND STREAMING—HOW AI CHOOSES WHAT YOU WATCH

There's something almost magical about turning on the TV these days and seeing a list of movies or shows that feel like they were picked just for you. That's no accident. Smart TVs and streaming services like Netflix, YouTube, and Amazon Prime use AI to learn your likes and dislikes. Each time you rate a movie, finish a series, or skip a particular genre, the system takes note. Over time, it starts suggesting programs that fit your taste—maybe British mysteries, travel documentaries, or uplifting comedies. Gone are the days of scrolling for half an hour to find something decent; now, your TV serves up a tailored menu. These recommendations aren't always perfect, but they get sharper as you use the service more. The more you watch, the better it understands your preferences.

The real fun begins when you use a voice remote or built-in assistant to control what's on screen. Instead of hunting for tiny buttons or squinting at channel lists, you can speak your request. Say you're in the mood for a British detective show. Press the microphone button on your remote and state clearly, "Show me British mysteries on Netflix." Your TV will search across multiple apps and bring up options within seconds. Feeling like music instead? Try "Play jazz on YouTube," and let the system do the work. Adjusting the volume is just as easy: "Turn up the volume," or "Mute." No need to fumble with the remote in the dark or shuffle through menus—voice control keeps things simple.

Setting up profiles is a game-changer, especially if you share your TV with others or want things just right for your eyes and ears. You can create a profile with your name, pick a big, readable font, and even choose a favorite color scheme. Many streaming apps allow you to increase text size or bold menus for better clarity. If hearing is a concern, enabling closed captions is straightforward. Dive into your TV or streaming app's settings and look for "Accessibility" or "Subtitles." Turn on captions, then tweak the font size until it feels comfortable. This way, you catch every word, even with background noise or if the actors have thick accents. For an extra layer of comfort, explore parental controls if you have grandchildren visiting or want to block certain content.

Minor frustrations can sneak in with smart TVs and streaming devices, but they're not insurmountable. Remotes tend to vanish between couch cushions just when you need them most. On some systems, you can ask your assistant to help: say "Find my remote," and if it has a beeping feature, it'll chirp until you spot it (on compatible remotes). If the screen suddenly freezes or goes black—a common headache—don't panic. Start by checking if the TV is plugged in and the power strip is on. If everything looks fine but nothing changes, try unplugging your streaming device (like Roku or Fire TV) from power for about ten seconds. Plug it back in and wait for it to restart; most glitches clear up with this trick. Occasionally, devices need a full reset. Look for a small button labeled "Reset" on the device's back or bottom—press gently with a paperclip according to instructions in the manual.

If navigating menus still feels like wading through molasses, use search features instead of scrolling through endless lists. Tap the microphone and say what you want—whether it's "Show me comedies" or "Go to channel 7." When all else fails and you hit a wall, nearly every streaming app has a help section in its menu. These often include large-print options or video tutorials designed for users who might not be tech experts.

Streaming platforms also let you create "watchlists" so your favorite shows are always easy to access. Add new series with one click—or one spoken command—and return later without hunting through menus again. If ever you see a message about needing an update or something not working as expected, don't ignore it: select "Update Now" if prompted, as these updates often improve both speed and accessibility.

The beauty of this modern setup is that it gives you more freedom over what you watch without extra hassle. Whether you're searching for old favorites or discovering new ones, AI quietly makes suggestions that fit your style. Voice features reduce stress on your hands and eyes. Accessibility settings make everything friendlier for every user. And if technology throws a curveball now and then—a lost remote, a frozen screen—there's almost always a simple fix close by. The remote-control struggle becomes less of a battle, and screen time becomes something to look forward to instead of dread.

"SAFE TO TRY" APPS—TRUST BADGES AND RECOMMENDED TOOLS

Choosing the right app can feel like picking a ripe fruit at the market—so many options, but some are better for you than others. To give you peace of mind, I've put together a shortlist of "Safe to Try" apps that really work for folks looking for simple solutions in health, communication, and everyday organization. Medisafe is a favorite among seniors who want a reliable medication tracker. It gently reminds you about each pill, lets you log what you've taken, and even allows a trusted family member to check in if you wish. For keeping your photos organized without headaches, Google Photos sorts snapshots by faces, dates, and places—and frees up space on your phone or tablet. Its search features are remarkable; type "birthday" or "dog," and old memories pop up instantly. WhatsApp is another gem, making calls or sending messages to loved ones a breeze, even across continents. It's secure, easy to use, and supports video calls, so you can see family faces anytime. Every app here comes with a "Safe to Try" badge because trusted groups, including senior advocates and health professionals have widely recommended them.

When browsing the app store—whether Apple App Store or Google Play—look for clues that an app can be trusted. Official apps always display their store's logo or badge. High ratings are another good sign; if other seniors say the app is easy and helpful, chances are it will be for you too. You might notice an "Editor's Choice" ribbon or a label like "AARP Recommended"—these aren't just marketing fluff but signals that the app has been reviewed for safety and usability. Stick with apps from well-known companies or those with lots of positive reviews from people in your age group.

Before you tap "Install," take a moment for a safety check. First, look at the app's permissions—it will ask if it can use things like your location, contacts, or camera. A medication tracker like Medisafe might need access to your calendar for reminders, but it shouldn't ask for your photos or microphone unless you want those features. If something feels off—say, a photo app wants access to your call logs—pause and double-check its purpose. Verify who made the app; developer names should match the company (for instance, "Medisafe Inc." or "Google LLC"). If you spot a strange developer name or lots of typos in the description, that's a red flag.

Getting set up is just the beginning. Here's a large-print checklist you can follow every time you try a new app:

1. Download only from your device's official app store.
2. Check the star rating (aim for four stars or higher).
3. Read one or two reviews from other seniors.
4. Before opening, review permissions: deny anything that doesn't make sense.
5. Verify the developer name is correct.
6. If anything makes you uneasy, skip the app and look for another with better reviews.

Sometimes an app turns out not to fit your needs—or worse, it asks for too much information. Don't let this shake your confidence. Deleting an app is quick and secure. On iPhones or iPads, press and hold the app icon until it wiggles, then tap the small "X" or "Remove App" button. On Android devices, press and drag the app icon into the "Uninstall" area at the top of your screen. If an app requests information that feels private—like your Social Security number or banking details—and you didn't expect it, close the app immediately and remove it using these same steps.

If an app locks up or starts acting strangely, try restarting your device first; many minor glitches disappear after a quick reboot. Should you ever worry that something isn't right—maybe the app sends spammy messages or changes settings without asking—delete it straight away. You can always ask a tech-savvy family member or friend to take a look if you're unsure.

Checklist: Safe App Setup

- Download only from the Apple App Store or Google Play
- Check star rating; aim for 4+ stars
- Read reviews from seniors
- Review permissions before opening
- Confirm the developer name matches the company
- Delete if unsure or uncomfortable

Using apps shouldn't feel risky. With these tips in mind, you can confidently explore new tools while protecting your privacy and peace of mind.

As we wrap up this chapter, you now have clear ways to spot safe apps and use them wisely in your daily life. These tools aren't just convenient—they're stepping stones to greater independence and connection. Up next, we'll explore how AI keeps you linked with friends and family, opening new doors for meaningful conversation and support no matter where life takes you.

CHAPTER 3
AI IN COMMUNICATION —STAYING CONNECTED WITH LOVED ONES

VIDEO CALLS MADE EASY—HOW AI HELPS WITH ZOOM, FACETIME, AND WHATSAPP

NOT LONG AGO, connecting with distant loved ones meant waiting for a letter or a phone call. Now, seeing grandkids' faces or talking with friends happens instantly from your home. This shift, while exciting, may also feel overwhelming. Fortunately, artificial intelligence has made video calls easier and more intuitive, even for those who don't consider themselves "techie." If new apps or worries about appearance have kept you away, you'll be glad to know how user-friendly things have become.

AI quietly simplifies video calls in ways you might not realize. For example, if you've ever felt distracted by your image or wished for better lighting, AI now automatically adjusts brightness, making faces clearer, even in dim or uneven lighting. It also filters out background noise, so your voice is crisp, and distractions like traffic or a TV don't interrupt the conversation.

A standout feature is face tracking. Apple's FaceTime, for example, offers "Center Stage," where AI keeps your face centered even as you move, so you don't fuss with camera angles or worry about staying still. Zoom and WhatsApp use AI to blur backgrounds. Whether you want to hide clutter, keep family photos private, or prefer a neater look, turning on background blur keeps you in focus while fading everything else, which enhances privacy and the overall quality of the call.

Getting started is easier than it sounds. To try Zoom for group chats or classes, head to your app store, search "Zoom," and download it for free. Once installed, open it and follow the simple steps to create an account. Adding a contact is simply entering their email, and joining a call involves tapping a link sent to you—no codes or repeated passwords required. To start your meeting, tap "New Meeting," then "Invite" and send invitations in just a few taps.

FaceTime works seamlessly with Apple devices. Open the FaceTime app, tap "+," add a contact from your saved numbers, then tap "Video" to call. WhatsApp is user-friendly on both smartphones and tablets: download, set up with your phone number, and tap the camera icon next to any contact for a video call. All major platforms support one-tap links, making joining or starting calls simple.

AI also improves accessibility. For example, Zoom allows live captions, appearing as on-screen text, so you can follow conversations if you have trouble hearing. Just look for "Live Transcript" during your Zoom call and activate captions with a tap. On FaceTime, use voice commands (e.g., "Hey Siri, FaceTime Barbara") to start calls without typing or searching—helpful if you find on-screen navigation tricky.

Safety and privacy are key concerns. If a call comes from an unknown number, tap "Decline" or close the window—no need to explain. If someone bothers you repeatedly, block them easily: go to their profile, press "Block Contact," and they can't reach you again. This ensures you control who can contact you.

Exercise: Your First Video Call Checklist

- Pick your app: Zoom, FaceTime, or WhatsApp
- Download from your device's app store
- Set up an account if needed
- Add a trusted contact

- Test the camera and microphone
- Try background blur for privacy
- Turn on live captions if needed
- Use voice commands to start calls (if available)
- Decline unknown calls
- Block anyone who makes you uncomfortable

Each small step brings greater confidence and the joy of connecting. Ask a friend or family member to join a practice call—they'll be happy to help you get started.

AI-POWERED MESSAGING—FROM AUTO-REPLIES TO TRANSLATION

If texting ever felt like a chore, you're not alone. Big buttons, small screens, and fumbling to find the right words can sap the fun from staying in touch. Thankfully, AI-powered messaging has changed the game. Now, your phone or tablet doesn't just sit and wait for your fingers to catch up—it tries to help. You may have noticed that when you start typing a message, little bubbles pop up with suggestions like "Sounds good!" or "On my way!" These are predictive replies, and they save time, especially when a quick answer is all you need—no more pecking at tiny keys or worrying if your reply is too short. The more you text, the more these suggestions fit your style, learning from your most-used phrases and offering them right when you're about to tap them out.

Texting can be even easier with voice-to-text tools. If your hands tire easily or the keypad feels too cramped, look for the small microphone icon in your messaging app. On both iPhones and Androids, tapping this icon lets you speak your message. Your phone listens and turns your words into text almost instantly. It's like having a patient secretary who writes down everything you say—no spelling worries, no racing against autocorrect. Just speak normally, pause for punctuation if you want, and check that the message looks right before sending. You'll be surprised at how accurate and fast it works, even with background noise or a softer voice.

Communication isn't limited by distance or even language barriers anymore. AI translation features are making it possible to chat with relatives or friends who prefer another language. On WhatsApp, for example, incoming messages can be translated with a simple tap—just press and hold a message, then select "Translate." Google Messages offers similar tools for text threads, allowing you to follow conversations in real time even if someone writes in Spanish, French, or another language. This is incredibly helpful if you have family abroad, or maybe new friends at a community center who feel more comfortable writing in their mother tongue. These translation features rely on advanced algorithms, but feel easy and natural to you as the user. You might not even realize AI is working behind the scenes, quietly bridging gaps that once seemed too wide.

Security remains a top priority in today's digital world. Keeping private conversations safe is more than just a comfort—it's a necessity. Many messaging apps now allow you to set messages to auto-delete after a specific period. In WhatsApp and Signal, you can choose this feature for sensitive chats; messages disappear after a day or a week, leaving no digital trace behind. This helps if you ever worry about leaving private details on your phone or tablet by accident. For even more control, apps let you lock specific chats with a PIN or fingerprint. Only you can open those conversations, so even if your device falls into the wrong hands, your information stays protected.

Deleting old messages is another smart move for privacy. In most apps, press and hold on any conversation and choose "Delete." If you want to keep things tidy but can't remember which chats need to go, AI can help here as well—some messaging platforms suggest which conversations are inactive or rarely used, making cleanup less of a guessing game.

Juggling conversations across different apps can get confusing fast, especially when family prefers WhatsApp but friends stick to regular texts. AI comes to the rescue by organizing your inbox so that important contacts rise to the top based on who you message most often. Your phone might suggest adding someone as a favorite if you talk regularly or highlight new messages from close friends first. This way, you don't have to scroll endlessly or worry about missing time-sensitive news.

If you ever misplace an important message or thread, search tools powered by AI make retrieval simple. Just type in a keyword—like "birthday" or "doctor"—and the app finds related messages instantly. No more scrolling through endless lists or opening the wrong chat by mistake.

For those hesitant about oversharing, privacy settings are easy to adjust. Open your messaging app's settings and look for controls like "Last Seen," "Read Receipts," or "Profile Privacy." You decide who can see when you were last online or read their message. On WhatsApp, you can limit this to only your contacts or turn it off altogether if you prefer more privacy.

When new technology feels complicated, remember you don't have to learn everything in one day. Try sending a voice-to-text message or translating a note from a friend who speaks another language. Set up auto-delete for one chat to see how it feels or rearrange your favorites list for easier access next time someone calls or texts. If something doesn't look right—maybe an unknown number texts late at night—just delete it and block the contact if needed. There's no harm in being cautious.

These features aren't just for tech wizards; they're built to make life simpler for everyone willing to give them a try. Whether it's sending birthday wishes across languages, replying quickly while out for a walk, or securing sensitive family news, AI-powered messaging brings peace of mind and makes connecting with loved ones feel natural again.

ORGANIZING CONTACTS AND GROUP CHATS WITH AI

Ever notice some names appear more often in your contact list? That's because modern phones and tablets use artificial intelligence to sort your contacts by how frequently you interact with them. Instead of scrolling through a long list, those you call, text, or video chat with most show up at the top. Your device automatically identifies who you talk to the most, making it much easier to find important people like your best friend, doctor, or family members. Most smartphones now have sections labeled "Favorites" or "Suggested," which use your call and message history to recommend contacts for easier access, streamlining your digital address book.

AI also identifies communication patterns, such as reaching out to your daughter every Sunday or texting your walking group on Thursdays, making those contacts easily accessible whenever you need them. Devices may even organize contacts by relationship—family, neighbors, coworkers, or friends in other cities—facilitating fast connections, especially if you add relationship labels (like "son" or "neighbor") when saving contacts. For friends in distant locations, AI uses location data to help you reach them quickly for travel plans or events.

Creating group chats is simpler than ever, thanks to these innovative tools. Want a group just for your grandchildren? Name it "Grandkids Club," pick a friendly icon, and add all the members in a few taps. WhatsApp makes this process especially easy: tap "New Group," select the right people, pick a name and icon, and you're ready to start sharing messages and photos with everyone at once. Similar groups can be set up for church friends, book clubs, or recipe-sharing neighbors, each with its group photo that's visible to all participants.

Managing groups is straightforward. If someone changes their number or moves away, you can easily add or remove them without starting a new group. By long-pressing on a group chat, you can rename, change icons, or mute notifications when the conversation gets too active at inconvenient times. AI helps keep active groups near the top of your chat list, moving less-used groups further down for a tidy inbox.

Messages can sometimes get lost in busy chat lists. Pinning meaningful conversations prevents this—just swipe right on a favorite contact or group and hit "Pin." This keeps essential chats like family or support groups at the top, so they're always easy to find, no matter how many messages come in.

Contact lists often accumulate duplicates or outdated numbers, especially when someone gets a new phone or is saved twice. AI-powered tools, like Google Contacts "merge duplicates," identify similar entries and let you combine them with one tap, helping you avoid confusion about which number is correct. Phones may also suggest archiving or deleting contacts you haven't used in a long time, keeping your list current and focused on those who matter.

Removing old or unused contacts is easy and helps keep things organized. Most contact apps include a "Suggestions" or "Cleanup" feature that recommends deleting numbers you haven't used for months or years. Removing contacts is safe—they disappear from your device, but are often recoverable from backup if needed. For extra peace of mind, export your contacts as a backup before cleaning up, which is usually found in the settings menu.

By letting AI organize your contacts and manage group chats, you can devote your attention to meaningful conversations and relationships, staying connected in a way that feels effortless rather than overwhelming.

Reflection: Personalize Your Digital Connections

Look over your contact list. Are there people you talk to every week but haven't marked as favorites? Pin them for fast access. Consider which groups—family, friends, neighbors—might benefit from a shared chat space. Try updating a group name or photo to make interactions more fun. You may find that having your contacts and chats reflect your real-world relationships makes communication more enjoyable and personal.

SMART EMAIL MANAGEMENT—HOW AI SORTS AND FLAGS YOUR MESSAGES

Opening your email inbox can sometimes feel like standing at the end of a hallway with dozens of doors, not knowing which ones hold friendly notes from loved ones and which lead to a pile of advertisements or scam offers. The good news is that you no longer have to sort it all out yourself. AI-powered email apps now do much of the heavy lifting, making sense of the chaos so you don't have to. Platforms like Gmail, Outlook, and Apple Mail use smart filters to organize messages as they arrive. Gmail splits your inbox into "Primary," "Social," and "Promotions" tabs, so important notes from family or your doctor don't get lost among store sales or social media updates. AI algorithms scan the subject, sender, and content, automatically sorting messages into these categories. You'll probably find that most of the time, legitimate personal mail lands right at the top of your "Primary" tab, while newsletters or coupons are shuffled off to "Promotions." This system saves time and helps you focus on what matters.

Setting up filters or flags can add extra order to your inbox, especially if you want to never miss a message from someone special. Marking an email as "important" teaches your app to prioritize it. In Gmail, click the star next to a message—this pins it for easy access later. Outlook and Apple Mail offer similar flagging options, letting you color-code or prioritize emails with a tap or click. Many apps allow you to set up notifications for messages from certain people only. For example, you can designate family members or close friends as VIPs. After adding them to your VIP list, your phone or tablet will alert you only when those folks send something new. This keeps notifications manageable and ensures you never overlook meaningful updates.

If you ever feel overwhelmed by unwanted emails or newsletters you don't remember signing up for, unsubscribing is easier than ever, thanks to AI tools built into most major email apps. Look for an "Unsubscribe" button near the top of the message or at the bottom in tiny print; clicking it removes you from future mailings. Gmail and Outlook often detect subscription emails and add an "Unsubscribe" link right next to the sender's name for quick removal. If spam sneaks through—emails about miracle cures or suspicious financial offers—flag them as spam with one click. AI learns from these actions, sharpening its ability to block similar messages in the future.

Accessibility features built into email apps also deserve attention. If you ever squint at small print or tire from scrolling, try activating large-text mode. On iPhones and iPads, go to Settings -> Display & Brightness -> Text Size, then drag the slider until text feels comfortable. Android phones offer similar adjustments under Settings -> Accessibility -> Font Size. Some email apps provide their own options to zoom in on text or switch to high-contrast themes for easier reading.

Writing emails can be just as easy as reading them, even if typing feels awkward. Dictation tools turn your spoken words into written text with surprising accuracy. On Apple devices, open a new email, tap the microphone icon on your keyboard, and start speaking; your device converts your voice into text instantly. The same goes for Android—tap the microphone button within your keyboard or in Google Assistant and dictate your reply. You can even tell your phone when to insert punctuation by saying "period" or "comma." If you make a mistake, correct it with a quick tap or by saying "delete last sentence." Voice dictation makes composing long replies or telling stories far less tiring.

Sometimes managing email can feel like a chore, but with AI's help, it becomes much more manageable—even pleasant. Filters and tabs mean fewer distractions and more time spent reading what's important. Flags and VIPs give you peace of mind that nothing crucial will slip by unnoticed. With unsubscribe links at your fingertips, your digital mailbox gets cleaner without effort. Accessibility features make reading easy on your eyes and hands, while dictation turns complicated replies into a breeze. When you use these tools together—filters keeping spam away, flags marking favorites, voice dictation writing your words—you'll find email becomes less of a daily headache and more like a helpful assistant in your pocket.

You might also notice that your app will prompt you with gentle reminders if it thinks you forgot to reply to something meaningful, nudging old but important conversations back up top so nothing slips away unnoticed. These little touches are powered by machine learning but feel surprisingly personal. Over time, your inbox starts working with you instead of against you—a small but meaningful win for staying connected in today's busy world.

SAFE SHARING—PROTECTING YOUR PRIVACY IN CONVERSATIONS

It's easy to feel comfortable in conversations—whether with friends, family, or new online acquaintances—and forget how important it is to safeguard your personal information. Sharing stories or recipes is harmless and great for building connections. But the digital world makes it easy to blur the line between safe and risky sharing. For example, if you get a message from someone—maybe calling you "Grandma" or "Grandpa"—asking for your Social Security number or bank account information, don't rush to respond, no matter how urgent or convincing it seems. Requests for sensitive details via text or email are a common tactic for scammers. Even if the person claims it's for something important, like a password reset or a relative in trouble, always pause and think. These urgent requests are classic attempts to trick you into giving up private details.

Treat online conversations as you would your mailbox: you wouldn't write personal details like your Social Security number on a postcard for anyone to see. Apply this mindset to your digital messages—only share private information with trusted people, via secure methods. If in doubt, call the person directly using a number you have already saved, not one provided in a suspicious message. This simple habit can spare you major problems later.

Understanding privacy settings in your favorite apps is another key to staying safe. Most messaging and video apps let you control who can see your activity, profile photo, or status. In WhatsApp, for example, open Settings -> Account -> Privacy. Here, you can restrict who views your profile photo, status, and "last seen." You might want only your contacts to see certain info, or set your status to private for more peace. On Facebook Messenger and similar apps, privacy or security sections typically let you hide your online presence or restrict who can contact you. Spending a few minutes to tailor these settings keeps you in control of your information.

Suspicious links are a frequent trap. You might receive an email asking you to "confirm bank details" or a text claiming you've won a prize and requesting your information. Phishing attempts can look surprisingly legitimate, sometimes using authentic company logos or names. If a message feels "off"—strange spelling, an odd sender, or unusual requests—trust your instincts and don't click. Delete the message or mark it as spam. If an official-sounding message asks for sensitive information, always visit the company's official website by typing the address yourself or call a verified number.

Blocking or reporting unwanted contacts is easier than ever, and a vital safeguard. If someone bombards you with messages or sends you something inappropriate, block them. Most messaging apps let you open the conversation and tap the name or number to find options like "Block Contact" or "Report Spam." This prevents future messages or calls from reaching you. Apps like Messenger, WhatsApp, and iMessage all have simple processes for reporting spam or harassment, helping keep everyone safer.

It's also wise to regularly check what permissions your apps have. Visit your phone's settings (under "Apps" or "Permissions") and review which apps can access your camera, microphone, or contacts. If an app's permission request doesn't make sense—like a weather app wanting your contacts—turn it off. You can always turn it back on if needed.

Make privacy checks part of your routine. Every few months, review who can see your status, clear old chats with sensitive info, and update your passwords if something feels off. Use strong, hard-to-guess passwords, not pet names or birthdays. If an app offers two-factor authentication—like a code sent to your phone—activate it for extra protection.

Sharing stories and laughs should feel enjoyable and safe. With some caution and practical habits, you can stay connected without worry. Be vigilant—lock your digital doors just like you'd lock your home at night. If a message seems suspicious, ignore it; if someone makes you uncomfortable, block them; if an app is too intrusive, stop using it. Using these habits, you can enjoy online interactions with confidence.

AVOIDING SCAMS—SPOTTING FAKE MESSAGES AND CALLS WITH AI'S HELP

Scams are getting sneakier every year, and these days, fraudsters don't just rely on the old tricks—they use technology too. Fortunately, you have some strong allies on your side now. Artificial intelligence isn't just about making your phone smarter or your emails neater; it's also a silent watchdog that helps protect you from tricksters looking to take advantage. Take call screening as an example. If you use a phone powered by Google Assistant, you might notice that unknown callers are screened before the call even rings through. The AI answers and politely asks the caller to state their name and reason for calling. You see a transcript of what they say before deciding whether to pick up, letting you avoid robocalls or scammers trying to catch you off guard. This simple feature means you get fewer interruptions from strangers and can focus on real calls from people who matter.

Email systems have also become much sharper. Services like Gmail have built-in AI that scans every message before it lands in your inbox, looking for telltale signs of trouble. If an email seems suspicious— maybe it claims your bank account was hacked or urges you to "act now" by clicking a strange link—the system might label it as "phishing detected." These warnings alert you to danger before you even open the message. Sometimes, the email gets sent straight to spam, so you never have to see it at all. Text messages are getting similar treatment. On Android phones, Google's spam detection flags suspicious texts and, in some cases, blocks them automatically.

It's also important to recognize the red flags you should watch for yourself. Scammers often target older adults with convincing stories that tug at your heartstrings or play on fear. Imagine getting a call or message claiming your grandchild is in trouble—maybe stranded in another city, needing money wired right away for bail or a hospital bill. This is called a "grandparent scam," and it's become more common than ever. The scammer might even know your grandchild's name, which can make the story feel all the more real. Another classic trick is fake tech support: someone calls claiming your computer is "infected" and offers to fix it if you give them remote access or pay for software you don't need. Any message that asks for urgent action, requests money quickly, or wants private information like bank details or passwords should give you pause. Always stop and think before responding.

AI tools aren't just for blocking scams—they also let you fight back by reporting suspicious activity. In WhatsApp or iMessage, hold down on the sketchy message or number, then look for options like "Report" or "Mark as Spam." When you flag something as unwanted, you not only block future messages from that sender but also help improve the AI for everyone else. Gmail offers a similar system; click the exclamation mark labeled "Report phishing" on any questionable email, and send it to Google's review team. This collective effort strengthens the safety net for all users.

To help keep things safe and straightforward, consider keeping a checklist handy near your phone or computer—a quick glance can remind you of smart habits before you answer or reply. Here's one you can print or write out:

Safe Communication Checklist

- Never share passwords over the phone, text, or email
- Always double-check requests for money, even from family
- Ignore urgent messages asking for personal info
- Verify unknown callers by calling back with a trusted number
- Don't click links from unknown senders
- Report and block suspicious messages or calls
- Ask someone you trust if something feels off

It's easy to let curiosity get the better of you when a message seems urgent or emotional, but taking those extra seconds to pause can make all the difference.

AI has become a quiet partner in your digital life—not just organizing and connecting but also shielding you from harm. By learning what red flags look like and using these built-in tools to filter and report trouble, you can enjoy the benefits of technology without unnecessary worry. The digital world isn't perfect, but with smart habits and a little support from AI, staying safe feels much less daunting.

As we wrap up this chapter, remember that connecting with family, friends, and your community should be enjoyable, not stressful or risky. Let technology do the heavy lifting when it comes to security. Up next, we'll look at how AI can help organize your daily life beyond communication, making routines smoother and freeing up more time for what truly matters.

CHAPTER 4
ORGANIZING YOUR DIGITAL WORLD— PHOTOS, FILES, AND REMINDERS

SORTING FAMILY PHOTOS WITH AI—EASY STEPS ON YOUR DEVICE

HAVE you ever opened your phone or tablet's photo gallery only to feel lost in a sea of snapshots—vacation sunsets blending into birthday cakes, grandkids' grins buried under a dozen pictures of your cat snoozing on the couch? It's easy for digital photos to pile up, especially when the camera is always just a tap away. The good news is, you don't need to become a computer whiz to bring order to that digital chaos. Thanks to artificial intelligence, your device can help you sort, organize, and even tidy up your photo collection, all with just a few gentle prompts.

Most modern smartphones and tablets come with built-in helpers you might not even know are there. Google Photos, found on many Android devices and available for iPhone too, has a handy "Assistant" that works quietly in the background. It spots patterns in your photos —like all those pictures from last July when you visited the beach, or every set of birthday candles blown out this year—and offers to group them into tidy albums. One morning, you might open Google Photos and find a ready-made collection called "Summer Vacation" or "Dad's 80th Birthday." These albums are AI's way of giving you shortcuts through your memories, so you can jump straight to your favorite moments without endless scrolling (see APA list #9).

Apple users have something similar with the "Memories" feature in the Photos app. Here, your iPhone or iPad will automatically gather images from special days—anniversaries, trips, or even just a sunny afternoon at the park—and arrange them into collections. It might surprise you with a slideshow set to music, reminding you of times you didn't even realize were special until you see them all together again (see APA list #30).

But what about those blurry snapshots or the four nearly identical takes of the family at Thanksgiving dinner? AI can help clear out the clutter by identifying duplicates and fuzzy images. In Google Photos, for example, the app will sometimes suggest deleting extra copies or photos that are out of focus. To do this yourself, open Google Photos and tap on "Suggestions." You'll see prompts like "Review duplicates" or "Delete blurry photos." Browse these suggestions, review the sets of similar pictures, and choose the ones you'd like to keep. With a few taps, you can tidy up dozens of images at once without having to inspect each one individually.

Let's walk through deleting duplicates step-by-step using Google Photos: First, open the app and look for the "Utilities" or "Suggestions" section. The app will show potential duplicates side by side. Tap on each duplicate to see it full size; then select "Delete" (usually marked with a trash can icon) on any unwanted copies. You don't have to worry about making a mistake—photos in Google Photos go into a trash folder first, so if you delete the wrong one, you can recover it within 30 days.

Once you've trimmed down your gallery, it's time to make sure your favorite images are easy to find. Both Google Photos and Apple Photos let you mark pictures as favorites by tapping a heart or star icon right below each photo. This small step creates a special album just for your favorite moments—no more hunting through endless folders to find that one perfect family portrait or the snapshot of your blooming garden from spring.

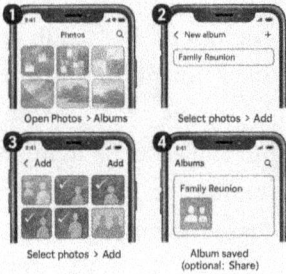
Create a New Album –
Google Photos

1 Open Photos > Albums
2 Select photos > Add
3 Select photos > Add
4 Album saved (optional: Share)

If you want a little more control, try creating folders or albums for specific people or events. Maybe you'd like one just for grandkids' art projects or another for your weekly book club gatherings. In Google Photos, tap "Albums," then "New Album," and give it a name. Add photos by selecting them and tapping "Add." On Apple devices, tap the plus sign in the Albums tab and select "New Album," then follow the same steps.

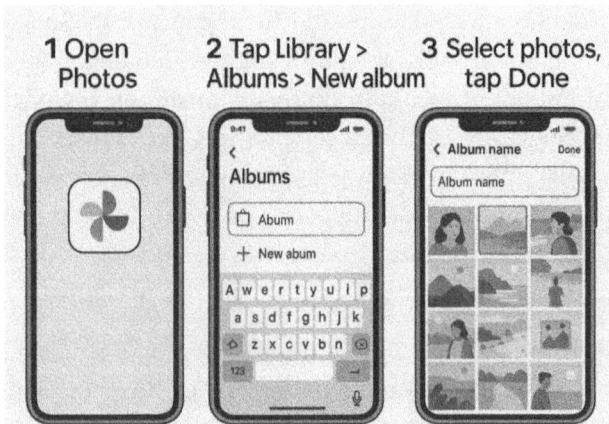
1 Open Photos
2 Tap Library > Albums > New album
3 Select photos, tap Done

Printable Checklist: Organizing This Month's Photos

Here's a simple checklist you can print out in large letters and keep near your device—a little nudge to stay ahead of digital clutter each month:

Photo Organization Checklist (Print Me!)

- Open your photo gallery or Google/Apple Photos app
- Review all new photos from this month
- Delete clearly blurry shots and accidental snaps
- Use "Suggestions" or "Utilities" to check for duplicates
- Tap the heart/star on your favorite images
- Create a new folder/album for grandkids or special events

- Move favorite photos into their new album
- Empty the trash/recycle bin after reviewing
- Enjoy looking back at your organized memories!

With these steps, organizing becomes less of a chore and more like reliving happy moments. You'll notice how quickly photo chaos gives way to neatly sorted memories—and how much easier it is to share them when family asks for that picture of last Sunday's barbecue.

FINDING PHOTOS FAST—FACE RECOGNITION AND TAGGING EXPLAINED

Scrolling endlessly through old pictures trying to find a favorite shot can be frustrating. Whether you're searching for every photo with your best friend or grandkids but can't remember where you saved them, face recognition powered by AI offers a simple solution. Your phone quietly recognizes faces that show up repeatedly and groups those images, so pictures of "Grandma Helen" or "Buddy the dog" are easily found. This means moments like your daughter's childhood can be pulled up in seconds rather than digging through endless camera rolls.

Most major devices use this feature. iPhones, for example, have a Photos app that uses face grouping to spot and collect frequent faces. You'll see small circles in the "People" album representing different faces; tap a circle to see every photo featuring that person, be it a recent picnic or a memory from years ago. Google Photos works much the same, recognizing and grouping faces even with changes in appearance like new haircuts or sunglasses. These face groups don't have names yet—you get to add them, making the process very easy.

Tagging is simply naming the faces, so searching for "Grandma Helen" or "Buddy" brings up every related image quickly. In Google Photos, open the "Search" tab and look for "People & Pets." Tap a face—person or pet—and use the "Add a name" option at the top. Enter any label, like "Grandma Helen" or "Buddy the dog," and save. On an iPhone, tap a face in the "People" album and select "Add Name" to do the same. Once tagged, simply type the name in the search bar and moments appear instantly.

Searching becomes even easier once you've tagged faces. If you're unsure what year a photo was taken but remember it was a Christmas in Florida, just type "Christmas" or "Florida 2019" in your device's search bar. AI-powered search sorts through your albums and brings up all matching photos—holiday dinners, palm trees, and more. In Google Photos, you can add more than one search term—try "Grandma Helen" and "Florida" to find every photo of her on that trip. It's almost like having a personal assistant who instantly remembers everything.

Naturally, many people get concerned about privacy with face recognition. The good news: face grouping on both Apple and Google devices runs locally or stays private in your account; it isn't shared with advertisers or outside parties (see APA list #3). You control who sees your photos, how they're labeled, and how they're organized. No one else can access these groups unless you choose to share them.

If you'd rather not use face recognition, turning it off is straightforward. On iPhone, go to Photos, tap your profile picture or initials, find "Photos Settings," then toggle off "Show People Names" or "People & Places." In Google Photos, tap your profile, choose "Photos settings," then "Group similar faces," and turn it off. Your photos remain safe—they just won't be grouped by face anymore.

Many find tagging to be more than just practical—it's enjoyable too. Giving labels feels like organizing old family albums, but now everything's digital and instantly accessible. Pets count as well, letting your cat or dog have their own collection of memories.

You can also search by places or events in addition to names. Type "beach" to pull up all sandy outings, or search "birthday" to view years of celebrations collected together. This works in both Google Photos and Apple Photos, requiring only simple taps and typing.

Don't stress about making mistakes. If you mistag someone or want to change a name, tap the edit button next to the label and update it. This flexibility lets you keep your library well-organized and always under your control.

In summary, face recognition and tagging are practical tools that save you time, helping you find your best moments fast. If you're concerned about privacy, you can easily manage face grouping settings for peace of mind, all right at your fingertips.

CREATING DIGITAL ALBUMS AND SLIDESHOWS WITH AI TOOLS

Sorting through hundreds of pictures on your phone can feel like searching for a needle in a haystack—especially when you want to pull together the best snapshots from a family reunion or a special trip. Thankfully, digital albums take the work out of gathering memories. AI helps by automatically offering groupings for you, but you can always take charge and design albums your way. In Google Photos, you'll notice suggestions pop up from time to time, like "Auto Album: Weekend at the Lake," where the app recognizes several photos taken at the same spot or over a weekend and puts them together for you. You just need to review and accept—no sorting required. If you prefer a more personal touch, you can build your own album from scratch. Tap the "Albums" tab, select "New Album," and start picking the photos you want included. Maybe you want an album just for "Family Reunions" or another titled "Spring Blooms." Add as many images as you like, and shuffle them into any order that feels right.

Digital albums don't have to stop at still pictures. Creating a slideshow is an excellent way to bring your memories to life and share them with friends or family. Most photo apps let you select a group of images and instantly turn them into a show, complete with gentle transitions and music. When you're in Google Photos or Apple Photos, look for options like "Create Movie" or "Slideshow." AI will suggest themes— maybe "Memories," "Summer Vacation," or even "Celebration"—that fit the mood of your pictures. Each theme adjusts the background colors, transitions, and sometimes even adds cheerful effects like confetti or falling leaves. You can pick from music included in the app or import your favorite tunes for that personal touch. The system will automatically time the photos to the beat of the chosen song, ensuring each moment gets its spotlight without dragging on too long.

Personalizing your album or slideshow is half the fun. Maybe you want your grandchild's first steps to show up before the birthday cake photo—or perhaps that group shot from last Thanksgiving deserves to close out the whole collection. Editing is simple: just drag and drop photos into your preferred sequence. Most apps let you tap and hold an image, then slide it left or right to rearrange. You can also add captions by selecting a photo and typing something meaningful under-neath—maybe "Grandpa's famous chili cook-off" or "First sunset in Florida." These little notes help tell the story behind each picture, so even years later, you'll remember exactly what made that moment special.

Special events deserve extra sparkle, and digital slideshows shine brightest when made for birthdays, anniversaries, or holidays. Imagine surprising your spouse with an "Anniversary Through the Years" slideshow—start by collecting one photo from each year of your marriage, then add them to a new album. Once organized, use the slideshow feature to lay out the sequence. You can overlay text directly on each photo—simply select "Add Text" within the editing tools and type something like "Our Wedding Day—1974" or "Alaska Cruise—2010." Don't be afraid to get creative with color choices and font styles; these options turn a simple project into a heartfelt keepsake.

For milestone birthdays, create a cheerful slideshow by gathering baby pictures, graduation shots, and recent family gatherings. Once orga-nized, pick lively music and sprinkle in captions such as "First Bike Ride—Age 6" or "Retirement Party." For anniversaries, consider starting with wedding photos and ending with current-day celebra-tions, weaving in messages like "Still dancing after all these years!" between images.

If you feel nervous about trying something new, remember there's no rush—most apps let you save your progress as you go, so you can come back later and add more pictures or edit text before sharing with others. If you ever want to preview your project before finalizing it, tap "Play" to watch how it unfolds on your device screen. This is a great time to adjust timing or swap out any photos that don't quite fit.

Interactive Project Prompt: Make Your First Slideshow

Ready to try it out? Pick a theme—maybe "Grandkids Growing Up" or "Adventures with Friends." Gather 10–20 photos that fit your idea. Open your favorite photo app and start a new album just for this project. Drag the photos into your desired order, add a few captions, pick out background music, and create your slideshow. Play it back for yourself first, then share it with someone special when you feel ready. You might be surprised at how easy—and rewarding—it is to see your memories come alive this way.

USING AI TO MANAGE NOTES, TO-DO LISTS, AND CALENDARS

Staying organized used to mean keeping a stack of notepads, a calendar on the wall, and maybe a pile of sticky notes on the kitchen table. Now, your phone or tablet can help you capture thoughts, plans, and reminders—no more lost scraps or forgotten appointments. Artificial intelligence quietly powers many of these tools, making it easier to manage everything in one spot, whether you're jotting down a grocery list or planning a doctor's visit.

Voice memos are a simple place to start. Maybe you get an idea while stirring soup or suddenly remember something important in the middle of the night. You can ask your phone or tablet to "take a note." Apps like Google Keep let you press the microphone icon and speak naturally. The AI listens, turns your speech into text, and saves the note instantly—no typing needed. It's quick, accurate, and forgiving if you stumble over words or speak slowly. If you prefer handwriting, some apps will even scan written notes using your device's camera. Apple Notes includes a "Scan Document" feature that lets you snap a photo of a handwritten recipe or letter. The app then turns it into a digital file you can search and store safely. No more shuffling through piles to find Grandma's pie recipe. These scanned notes are easy to organize and always at your fingertips.

To-do lists can get unruly fast, especially when you juggle errands, appointments, and hobbies. AI steps in to streamline the mess. Apps like Microsoft To Do help you create lists with just a few taps. You can add tasks, set deadlines, and flag items that need your attention soon. For example, if you need to refill a prescription every month or schedule a haircut by Friday, you just add it to your list and set a due date. As soon as you finish something, check it off—the app keeps track and sometimes even gives you a small cheer for completing your goals. If you're more visual, color-coding brings order to chaos. Assign blue to health tasks, green for groceries, red for bills—whatever feels right. This way, a quick glance tells you exactly what's urgent and what can wait.

Prioritizing isn't just about marking what matters most; it's about letting AI help you spot patterns. Many apps will start suggesting which tasks are overdue or need rescheduling based on your habits. If you tend to pay bills on the first Monday of each month or like to water plants every Saturday, the app notices and offers gentle reminders at just the right time. Some even propose new tasks before you realize you need them, like picking up milk if it's been a while since groceries appeared on your list.

Managing appointments used to mean flipping through paper calendars or writing tiny notes in the margins. Now, syncing your digital calendar across devices brings everything together without fuss. Link your Google or Apple calendar with your phone, tablet, and even computer—one account keeps everything updated everywhere. Add an appointment in the doctor's office on your phone, and it shows up automatically on your tablet at home. No more double-booking or wondering if you wrote something down twice.

AI makes scheduling smarter than ever before. If you receive an email confirming a dentist appointment or a lunch reservation, many calendar apps will automatically create an event without you lifting a finger. Your device reads the details—date, time, location—and adds them to your calendar. You'll often get a notification asking if you want to save the event; say yes, and it's set. Some apps also look at your existing schedule and suggest good times for new activities based on open spaces in your day—handy if you want to avoid conflicts.

Recurring reminders are another area where AI shines. If you have a regular book club, exercise class, or medication schedule, set up the event once and tell the app how often it repeats—daily, weekly, monthly—it takes care of the rest. You'll get reminders before each event so nothing sneaks up on you.

Smart suggestions are where things start to feel almost magical. For example, after attending a monthly bridge group several times, your app might suggest "Bridge Club" as soon as you begin typing "B." Or if you always call your daughter on Sundays at 7 p.m., next time she texts, your phone may prompt with "Set call with Mary for Sunday?" This gentle guidance helps you stay on top of routines without constant effort.

Connecting all these tools is simple once you're set up with one main account, whether Google or Apple. It keeps notes, lists, and appointments in sync across each device you own. If something changes—a new time for lunch or an added prescription refill—you update it once, and every device reflects the change.

For those who struggle with typing or remembering passwords, many apps now offer voice logins or simple fingerprint access. This little touch removes barriers and makes daily organization feel less like a chore and more like second nature. With these smart helpers in place, your days feel less rushed and scattered. The best part? You still stay in control, choosing what matters most and letting technology take care of the details behind the scenes.

VOICE-ACTIVATED REMINDERS—NEVER MISS AN APPOINTMENT AGAIN

Having a voice at your fingertips—or rather, ready to respond in your living room or pocket—can make life a whole lot smoother. Gone are the days of sticky notes falling off the fridge or calendars buried under a pile of mail. With digital helpers like Alexa, Google Assistant, and Siri, you can now set reminders for just about anything, using nothing but your voice. You might be surprised at how straightforward this feels. Imagine you're finishing up dinner and suddenly remember you need to take your evening medicine. Instead of hunting down a pen, you say, "Alexa, remind me to take my medicine at 8 p.m." Within seconds, Alexa confirms, "Okay, I'll remind you at 8 p.m." For those with an iPhone or iPad, calling out, "Hey Siri, remind me to call my daughter tomorrow at 10 a.m.," gets the same result. Google Assistant, often built into Android devices, is just as handy. A simple, "Hey Google, remind me to water the plants every Saturday morning," and you're all set.

Managing these reminders is just as simple as setting them. If you want to check what's coming up, ask your device directly—"Alexa, what are my reminders today?" or "Hey Google, what reminders do I have?" If something's changed and you want to edit or cancel a reminder, try saying, "Cancel my 8 p.m. medicine reminder," or "Change my reminder to call the doctor from Monday to Tuesday." For those who prefer using a screen, opening the Reminders or Assistant app on your device will show a list of all current reminders. Tap on any item to edit the time, change the description, or delete it altogether. It's all about flexibility—your reminders can be adjusted as easily as speaking or tapping.

Recurring reminders are perfect for routines that repeat. You might want a nudge for weekly book club meetings or monthly bill payments. To set these up by voice, say, "Alexa, remind me to pay the electric bill on the first of every month at 9 a.m.," or "Hey Siri, remind me to take out the trash every Thursday night." Google Assistant understands requests like, "Remind me every Friday to check the smoke detector batteries." When you use these commands, your device will automatically repeat the reminder at the right intervals until you tell it to stop—no need to reset anything week after week.

Sometimes reminders don't ring through as expected, and there can be a few reasons for that. The most common culprit is volume. If you miss a reminder or only see it flash by on screen, check if your device's sound is muted or turned down low. Smart speakers usually have a physical button to mute or unmute the microphone and speaker; make sure both are active. On smartphones and tablets, use the physical volume buttons and double-check if your device is in "Do Not Disturb" mode—a setting that silences notifications. Next, review your notification settings in the device's main menu. Look for options related to Reminders or Alarms and make sure they're allowed to send alerts and sounds. If needed, turn on banner notifications or lock screen alerts so you won't miss an important nudge.

If reminders still aren't coming through correctly, restarting your device can sometimes fix minor hiccups. Power it off and back on. This refresh can clear out glitches that might be blocking sound or notifications. Also, ensure your device is connected to Wi-Fi or mobile data—some assistants need an internet connection to function fully. If all else fails and reminders keep slipping by, try updating your device's software or asking a tech-savvy friend for help. Even a quick search online for your specific device's reminder troubleshooting can point you in the right direction.

Once you're comfortable with reminders, it opens up new possibilities for organization. Set gentle nudges for everything from medication schedules and recurring appointments to watering houseplants and calling old friends each Sunday. If you want more nuance—like a different sound for medical reminders versus household chores—some devices let you assign unique tones or even display reminders visually on smart screens around the house. For those who like backup plans, many devices allow reminders to show up as both audio alerts and on-screen messages; this dual approach helps ensure nothing falls through the cracks.

Voice-activated reminders put you in control without needing complicated steps. You can set them on a whim while doing everyday tasks—cooking dinner, folding laundry, working in the garden—and they'll faithfully remember what needs attention even if you forget. There's real peace of mind in knowing something is keeping track alongside you without fuss or drama. And if ever things seem off—a reminder doesn't appear or sound as expected—a quick check of volume settings and notification permissions usually solves it. Technology should adapt to your life, not the other way around. With these digital helpers in place, those little tasks and important moments get done right on time, leaving more space for everything else that matters most to you.

PRINTING AND SHARING MEMORIES—SAFELY SENDING PHOTOS TO LOVED ONES

There's something special about holding a printed photo in your hand, whether it's a family portrait or a snapshot from last summer's back-yard barbecue. Even with all the digital albums and screens, many of us still enjoy sending real photos to friends or putting them in a frame at home. The good news is, you don't need a fancy printer or a tech degree to turn those digital snapshots into something you can touch. If you have a home printer, start by opening your photo on your phone, tablet, or computer. Look for the "Share" or "Print" button—usually shown as a little square with an arrow or a printer icon. Select your printer from the list (ensure it's turned on and connected to Wi-Fi), preview the photo's appearance, and then tap "Print." For best results, use glossy photo paper and check the settings for print quality before starting. If you'd rather order prints the easy way, free apps from Walgreens, Shutterfly, or CVS can help. Download the app, choose your favorite photos, select print sizes, and pick either home delivery or local store pickup. These services guide you through every step, even letting you crop or adjust images so everyone's face is front and center.

Sharing photos online opens up another set of choices—and it also brings up questions about privacy and control. Sending single photos to family is simple with text messages or email. On an iPhone or iPad, select the picture, tap the share icon, then pick "Message" or "Mail." Add your loved one's contact and send. Google Photos offers a different approach: you can create a shareable link that only trusted people can see. Select the album or image, choose "Share," then "Create link." Paste this link in a message or email to your recipient. For extra safety, adjust the settings so only specific people with the link can access your photos—no public posting required. Remember, sending one photo at a time can help you avoid mistakes like sharing the wrong image or accidentally including something too personal.

If you're sharing a bunch of pictures at once—like from a special event or vacation—it's smart to use album sharing features where you control who gets in. Google Photos and Apple Photos both let you invite people to view an entire album by entering their email addresses or phone numbers. They'll receive a direct invitation and won't be able to see anything else in your library. Sometimes plans change, or you want to stop sharing; removing someone is as easy as opening the shared album, tapping on their name, and selecting "Remove Access." In Apple Photos, choose the album, tap the person's icon, and pick who should no longer see those images. This gives you peace of mind if you ever send something by mistake or want to limit access later on.

Privacy deserves your attention whenever you share personal memories. It's easy to get swept up in excitement and forget about safety. AI Demystified for Seniors: An Easy Beginner's Guide to Learn How AI and Smart Devices Review your privacy settings—most apps let you decide if recipients can download, share further, or just view your pictures within the app itself. Be cautious about sharing location data embedded in photos. Many smartphones automatically tag where a picture was taken; this information can be removed in the photo's details or settings before sharing if you'd rather keep your whereabouts private. When using services like Google Photos links or shared albums, always double-check the list of invited viewers.

Safe Sharing Checklist (Print This for Quick Reference)

- Confirm the recipient's name and contact information before sending
- Review all photos for anything private or sensitive
- Check privacy settings on albums or links
- Remove location data if not needed
- Avoid sharing large albums with people you barely know
- After sharing, review who has access and remove anyone you don't recognize
- Trust your instincts—if something feels off, pause before sending

Taking a few minutes to review these steps will help protect your memories and keep sharing joyful moments rather than stressful ones.

As you become more comfortable printing and sharing photos, these moments become easier to pass along—whether it's mailing a birthday snapshot to an old friend or texting today's lunch photo to your grandchild in college. These small actions build bridges between generations and help everyone stay connected.

Looking ahead, now that your digital life is neater and sharing feels safer, it's time to explore how AI-powered tools can support your health and wellness goals, turning reminders and smart habits into daily routines that help you thrive in this next chapter.

MAKE A DIFFERENCE WITH YOUR REVIEW
UNLOCK THE POWER OF GENEROSITY

"We can't help everyone, but everyone can help someone."

RONALD REAGAN

People who give without expecting anything back often find more joy in their own lives. A review is one of those small gifts that can make a big difference.

Would you help someone just like you—curious about **AI** but unsure where to start?

My mission is simple: **make AI easy, friendly, and useful** for anyone who wants to stay connected in today's digital world—without the tech headaches.

But to reach more people, I need your help.

Most people choose books based on reviews. Your honest words could be the friendly nudge a fellow beginner needs to take their first step.

It costs nothing and takes less than a minute, but it could change someone's journey with AI.

Your review could help...

- ...one more grandparent video chat with family across the country.
- ...one more retiree feel confident using a smart speaker.
- ...one more person discover how AI can make daily life easier.
- ...one more reader feel that technology is something they can handle.
- ...one more smile happen because someone stayed connected.

To make a difference, just scan the QR code below or visit this link:

**[https://www.amazon.com/review/create-review?&asin=
B0G3WVPNY8]**

If you enjoy helping others, you're my kind of person.

Thank you from the bottom of my heart!
— *Gwen Blake*

CHAPTER 5
HEALTH, WELLNESS, AND SAFETY—AI AS YOUR PERSONAL ASSISTANT

SETTING UP MEDICATION REMINDERS WITH AI HEALTH APPS

HAVE you ever stared at a row of pill bottles and wondered, "Did I already take these?" It's a familiar moment for many, especially when routines get disrupted or a new prescription appears. Remembering medications shouldn't feel like a memory game, and that's where AI-powered health apps can make a real difference. These smart tools act like a personal aide who never gets distracted, always ready to remind you at just the right time. If you've ever wished for a gentle nudge or some extra assurance, you're in good company. Today's technology lets you leave sticky notes and guesswork behind, making medication management smoother and less stressful than ever.

Choosing the right medication reminder app doesn't require any technical expertise. The best options are made with simplicity and safety in mind, perfect for seniors who want peace of mind without frustration. Three highly regarded choices are Medisafe, CareZone, and Apple Health. Medisafe stands out for its easy interface and large, readable text. After downloading from your phone's app store, you'll find straightforward menus with step-by-step guidance. CareZone offers similar benefits, focusing on clarity and gentle reminders while also allowing you to keep notes and even scan pill bottle labels. Apple

Health, built right into iPhones, lets you track medications alongside other health information, so everything stays in one place. No matter which you pick, you'll want to start with an app that feels comfortable —big buttons, clear language, and no unwanted surprises.

Once you've installed your chosen app, setting up your medication schedule is as easy as following a guided recipe. Start by opening the app and looking for an "Add Medication" or "Get Started" button. The app will prompt you to type in the name of your medicine—if spelling is tricky, most apps suggest choices as you type or let you scan the bar code right off the bottle. After selecting your medication, enter the dose (such as "one tablet" or "two drops") and the time of day you need to take it. You can add as many medications as you have; the app keeps them all organized in a simple list. For daily prescriptions, choose "every day," but if your schedule varies—like "Mondays and Thursdays only" or "every other day"—just select the appropriate option. The app also allows for "as-needed" reminders for occasional medications.

Customizing reminders is one of the best features for anyone with hearing or vision changes. Most apps let you pick an alert sound from a list—some chime gently, others are louder or more persistent. Adjust the volume in your phone's settings or within the app itself so you never miss an alert. If you prefer vibrations or visual cues (like a pop-up on your screen), turn those on in the notification settings. Each reminder appears right when it's needed, displaying the medication's name and dose so there's no confusion.

When an alert arrives, you'll see options to mark the dose as "taken" or to "snooze" if you're not ready. Tapping "taken" checks it off your daily list, giving a little sense of satisfaction and certainty. If life gets busy, use the "snooze" feature to delay the reminder—maybe for 10 or 15 minutes—so it returns when you're free again. These interactive notifications create a safety net without nagging.

Sometimes, sharing your medication schedule with a family member or caregiver brings extra peace of mind. Medisafe and CareZone both allow you to invite trusted people into your account. Using their email address or phone number, send an invitation so they can view your schedule and receive alerts if a dose is missed (with your permission). This feature is especially helpful if someone helps manage your care or if you want backup support on busy days.

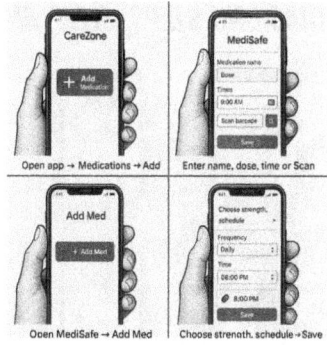

Apple Health takes a different approach by letting you print or email your entire medication list directly from the app. This makes doctor visits easier—you can hand over an up-to-date printout or send it ahead by email, saving time and avoiding confusion about prescriptions and dosages.

Interactive Exercise: Try Setting a Reminder

Pick one medication you take regularly—even if it's just a vitamin—and walk through these steps using Medisafe or CareZone:

1. Open the app and tap "Add Medication."
2. Type or scan the medication name.
3. Enter dose and preferred time.
4. Choose your alert sound.
5. Save and wait for the next scheduled reminder.
6. When it rings, tap "taken" or use "snooze" if needed.
7. Invite a family member if you'd like added support.

If you need to adjust times or doses later, simply edit the entry—no need to start over. As your needs change, add or remove medications with just a few taps.

These AI-powered tools don't just help with memory; they build a sense of control and safety around your health routines. Whether you're managing several prescriptions or want to keep track of occasional supplements, everything stays organized—and you stay confident that nothing will fall through the cracks (GoodRx Health, n.d., APA list #11).

TRACKING STEPS, SLEEP, AND EXERCISE WITH WEARABLE AI

1. Open App Store → search "Fitbit" /„Garmin Connect" (or open "Health" app)

2. Install & open → Sign in / Allow Bluetooth

3. Pair device → enter code on iPhone

4. Enable permissions & Health sharing → Done

Suppose you've ever wondered about your actual daily steps or wanted insight into restless nights. In that case, wearable fitness trackers and smartwatches like Fitbit, Apple Watch, and Garmin make these answers easy to access. Modern trackers have evolved far beyond pedometers—they're now user-friendly health companions powered by artificial intelligence to monitor movement, analyze sleep, and support healthier habits. These devices suit nearly everyone, including seniors, and don't require any tech expertise. Most people find them comfortable and simple after an initial setup.

Typically, getting started just means strapping the tracker on your wrist. Adjustable bands ensure a snug but comfortable fit. If the material feels rough, try swapping it for a softer silicone or fabric band, often available in stores for sensitive skin.

To connect your tracker, enable Bluetooth on your smartphone or tablet, open the companion app (like Fitbit, Apple Health, or Garmin Connect), and follow the pairing instructions. The app finds your device automatically—just confirm pairing with a tap or code.

Adjust accessibility settings for easier use: increase font size, tweak color contrast, and switch audio alerts to gentle vibrations if you prefer. Many trackers let you personalize displays, keeping key stats like steps or heart rate always within easy reach.

Once synced, real numbers start appearing. Each step counts toward a daily goal—usually set to 10,000 steps by default, but fully adjustable to fit your life or doctor's advice. Your tracker keeps an up-to-date tally; in the app, you can review your weekly activity and easily spot patterns with simple line graphs and color-coded bars. For instance, you might find Mondays are more active than Thursdays.

Sleep tracking is another helpful feature—wear your tracker overnight. It monitors movement and heart rate to estimate your time spent in light, deep, and REM sleep. In the morning, check a simple sleep score and see a chart breaking down your sleep stages. If you feel groggy but your device shows minimal deep sleep, try going to bed earlier or darkening your room; even minor improvements add up.

Most trackers also display weekly exercise summaries. They log not just formal workouts, but also walks, gardening, or cycling, recognizing various activity types. Some start automatically with motion; others require a tap to begin tracking. Watching your activity streaks grow can be motivating, even if your "workout" is just a stroll around the block.

Reminders to move are a favorite feature for gentle motivation. Set your tracker to nudge you hourly if you haven't been active—just a buzz or message like "Let's stretch our legs!" These prompts encourage easy movement: march during commercials, walk around while waiting for something to finish, or circle the living room between chapters of your book. If you skip them, don't stress—they're meant to help, not nag.

Trackers celebrate progress, too. Reach your step goal or sleep score, and the app sends upbeat messages or digital badges—sometimes shareable with friends or just for your own satisfaction.

Wearable AI adapts to your needs: lower goals if they feel overwhelming or reduce notifications if they're distracting. This is your tool, and you should use its data as a gentle guide rather than strict rules.

Staying active and getting restful sleep are foundational to wellness at any age. Wearable trackers reveal helpful patterns you might otherwise miss, making daily routines more manageable and even fun. With just a wristband and some curiosity, you can find new possibilities each day.

USING AI FOR DOCTOR VISITS—TELEHEALTH TIPS FOR SENIORS

You might remember when visiting the doctor meant arranging a ride, waiting in a crowded lobby, and filling out the same forms over and over. Now, with telehealth, you can talk to your healthcare provider right from your favorite chair. New telehealth services—like MyChart, Teladoc, and Amwell—take it a step further by using AI to make the whole experience smoother. These platforms aren't just a substitute for traditional visits; they can make medical care more comfortable, organized, and even safer, especially when getting out isn't easy or you simply prefer staying home.

The first step is signing up. Downloading the app is usually as simple as opening your phone or tablet's app store and searching for "MyChart," "Teladoc," or "Amwell." Once installed, launch the app and select "Create Account" or "Sign Up." You'll need to enter some basic information —your name, date of birth, and possibly your insurance details. The

app will likely ask you to create a secure password; pick something memorable but hard for others to guess. Some apps even suggest a strong password for you. If you frequently forget passwords, consider writing them down in a safe place or using a password manager built into your device.

Booking an appointment feels similar to making a dinner reservation online. Tap the button that says "Book Visit" or "Schedule Appointment." Select the type of doctor you want to see—primary care, specialist, or even behavioral health. Many telehealth platforms show you available times, so you can pick a slot that fits your schedule. After confirming your spot, the app will send a reminder by text, email, or on-screen notification. Some even add the appointment to your phone's calendar automatically. You don't have to worry about remembering; AI takes care of that part.

Joining your virtual appointment is straightforward. About ten minutes before your visit, open the app and look for a button labeled "Join Visit" or "Start Call." Some systems use a "virtual waiting room," where you'll see a message letting you know the doctor will join shortly. If you're early, you can relax; there's no need to wait in silence. Many platforms display helpful tips or checklists while you wait. When the doctor arrives, your screen will switch to a live video, like FaceTime or Zoom.

AI features work quietly in the background to help before and after your appointment. For example, many apps offer a digital symptom checker. This tool asks simple questions about how you're feeling—think of it as an interactive checklist. Your answers help the doctor prepare by flagging possible issues right away. You might be prompted to list any pain, coughing, changes in appetite, or medication side effects. This information is shared securely with your provider to ensure nothing important is overlooked.

Medication lists and appointment reminders also come into play. AI keeps track of your prescriptions and sends nudges when it's time for a refill or follow-up visit. After your appointment finishes, many platforms automatically send a summary of what was discussed, instructions for next steps, and reminders for future care—all in clear, plain language.

Technology isn't perfect, though. Sometimes video calls freeze or drop unexpectedly. If this happens during your visit, don't panic. Look for a tech support button—usually marked with a question mark or labeled "Help." Tapping this connects you with someone who can guide you through restarting the call or fixing sound issues. If audio drops but video remains, try toggling your microphone off and back on; it often solves minor glitches. In rare cases where nothing works, close the app completely and reopen it. The virtual waiting room usually lets you rejoin without starting from scratch.

Privacy is front and center in telehealth apps. Before your call begins, check your audio and video settings. The app lets you preview yourself on camera—if you see something you don't want to share (like a pile of laundry), just adjust your camera angle or use a virtual background if offered. You can also mute your microphone when not speaking for added comfort.

If you ever worry about privacy during the visit—maybe because of background noise at home or concerns about who can hear—most apps provide simple tools to mute audio temporarily or even pause video without disconnecting from the appointment. It's also wise to review the app's privacy settings before your first visit; look for options that limit data sharing outside your healthcare team.

Telehealth platforms are designed with support in mind. If anything feels confusing or stressful, reach out through built-in help features or ask your provider's office for assistance beforehand. Many offices have staff trained specifically to support seniors using these new tools, so don't hesitate to request help.

Virtual doctor visits with AI support aren't just about convenience—they're about giving you more control over your health without extra hassle. Whether it's scheduling an appointment in minutes, getting reminders just when needed, or having step-by-step support if something goes wrong, these tools are like having a personal assistant in your pocket—one who never forgets and always has time for your concerns.

MANAGING MEDICAL RECORDS SECURELY WITH AI TOOLS

Keeping track of personal medical records used to mean endless folders, loose papers, and the occasional scramble before a doctor's visit. These days, AI-powered health portals and apps put all your important information at your fingertips, securely stored and neatly organized. Tools like Apple Health Records, MyChart, and HealtheLife are designed to make life easier, not harder. They let you view test results, prescription history, and vaccination records in one place, with a few taps on your phone or tablet. No more rummaging through drawers or waiting for office staff to fax paperwork. Everything is private, but available when you need it most

Getting started is straightforward. First, download your preferred app —Apple Health Records is built into most iPhones, while MyChart and

Step-by-step: Apple Health Records, MyChart, & HealtheLife

1. Open App Store > search
2. Install & sign in; allow permissions
3. Link your clinic & enable Health
4. View & share records securely

HealtheLife work on both Apple and Android devices. Once installed, you'll need to create an account using your full name, date of birth, and sometimes a patient ID from your healthcare provider. These apps connect securely to your doctor's office or hospital system, pulling in information automatically. Some allow you to add details yourself if you have a paper record from another provider. For example, if you receive a printed lab report, use your phone's camera within the app to scan and upload it. The app converts the image into a digital file and files it under a clear label, such as "Blood Test April 2024." You can even add notes or tag the document for easy searching later.

Locating past results is just as simple. Open the app and select the "Records" or "History" tab. There you'll find categories like lab results, imaging, immunizations, and prescriptions. Want to check when you last got your flu shot? Tap "Immunizations," and the date appears right away. Curious about cholesterol trends over time? Scan the list of blood test results, often color-coded or displayed with simple charts for easy comparison. Prescription histories show what you've taken and when refills are due—a real help for conversations with doctors or pharmacists.

Sometimes you need to share part of your record with someone else—a new specialist, a family caregiver, or a physical therapist. These apps make it easy while still protecting your privacy. Instead of printing everything or handing over your whole history, you can generate a secure link that gives limited, read-only access to just the records you choose. Tap "Share," select the specific files or test results, and set an expiration—maybe one week or one month. This link can be emailed directly to your doctor's office or printed as a QR code for in-person visits. Once time runs out, access closes automatically, so you stay in control of your personal data.

Security matters more than ever when dealing with health records. The best apps use strong password protection and offer two-factor authentication for another layer of safety. When creating a password for your medical app, choose a long, complex password that includes a mix of letters (both upper and lower case), numbers, and special symbols like "$" or "!" Avoid using birthdays or common words. For example, instead of "Sarah123," try something unique like "Lily!Sunset47." Write it down in a secure spot if needed.

Two-factor authentication adds extra protection by asking you to verify your identity each time you log in from a new device. The app sends a code to your phone or email; you enter this before gaining access. It's a simple way to block unwanted intruders—even if someone guesses your password, they can't get far without that code.

Recognizing official health platforms is important, too, since there are copycat apps out there trying to trick people into sharing private info. Always download apps from your device's official store (Apple App Store or Google Play). Look for apps published by major hospitals, health networks, or companies you trust—never by unfamiliar developers with odd names or no reviews. Official health portals will explain how they protect your information and use clear language about privacy policies.

Visual Checklist: Smart Steps for Secure Health Records

- Download health apps only from official stores
- Set a strong password: upper/lowercase letters, numbers, symbols
- Enable two-factor authentication
- Scan/upload only from trusted providers
- Use sharing features that limit access by time and document
- Double-check the sender or app publisher before logging in

- Review the privacy policy for any app handling medical data

With these steps in place, managing medical records becomes less of a worry. You gain not just convenience but also peace of mind, knowing sensitive details are locked up tight but ready when you truly need them.

SPOTTING HEALTH SCAMS—AI'S ROLE IN SCAM ALERTS AND SAFE BROWSING

The internet is packed with health information, but sometimes it feels like walking through a crowded market where not every vendor is honest. For seniors, scams often hide behind friendly faces and promises that sound almost too good to ignore. You might see an email about a "miracle arthritis cure" or a website selling cheap prescription drugs with no doctor's visit required. These kinds of messages prey on hope and curiosity, especially when you're looking for relief or answers. Unfortunately, health scams have become more creative and convincing, using fake celebrity endorsements, phony approval stamps, and urgent warnings to pressure you into acting fast.

AI tools now play a crucial role in protecting you from these digital traps. When you check your email, most modern apps—like Gmail or Outlook—scan messages before you even open them. If an email comes from an unknown source or contains suspicious language, the app flags it as spam or marks it with a warning banner. For instance, messages promising instant results, demanding payment by gift card, or urging you to "act now" often get filtered out. Browsers like Google Chrome include Safe Browsing technology that checks websites in real time. If you click a link to an unlicensed pharmacy or a site promoting questionable supplements, you might see a big red warning page telling you to turn back. These AI-powered shields constantly learn from new scams, adapting to spot the latest tricks before they reach you.

Fake pharmacies are everywhere online and can be hard to spot at first glance. They may use real-sounding names like "USA Discount Meds" or display logos meant to look like government seals. But if a site says you don't need a prescription for controlled medicines, or offers prices that seem unbelievably low, those are big red flags. AI systems compare these details to known scam databases and often block access automatically. Similarly, miracle cure ads for things like COVID-19, cancer, or chronic pain rely on emotional language—"secret ingredient," "guaranteed results," "limited supply." Email filters look for these patterns, moving them out of your inbox to keep your focus on genuine health updates.

One of the smartest things you can do is pay attention to scam alerts and warnings when they pop up. If your browser flashes a red warning window before loading a health site, don't ignore it—even if the offer looks tempting. Close the tab and try searching for the product or company using trusted sources. Official health organizations like the CDC or Medicare have websites ending in .gov; these are safe places to check information before making decisions. Many email apps let you report suspicious content with a single click—look for buttons labeled "Report Spam" or "Report Phishing." Reporting doesn't just protect you; it also helps AI systems learn and keeps others safer.

To make things even easier, here's a quick checklist of red flags that signal a likely scam: if you see phrases like "No doctor's prescription needed," "Pay with gift card," "Secret cure doctors don't want you to know," or messages that create panic—"Limited supply, act now!"— step back and think twice. Other warning signs include requests for personal information (like Social Security numbers or complete credit card details) in emails or pop-ups, misspelled words, blurry logos, or links that look strange when you hover over them with your mouse.

If you ever feel uneasy about a message, don't feel pressured to respond. Instead, block the sender using your email app's settings. Most apps have simple menus for blocking unwanted contacts— usually under "More" or "Settings." If you've already replied or shared information and something feels wrong, reach out to a trusted friend or family member for help right away.

Whenever you need reliable health advice online, stick to official resources like the CDC (cdc.gov), Medicare (medicare.gov), or your local hospital's website. Avoid clicking on links sent by unknown contacts and never download attachments from emails about health products unless you expect them. Many scammers use attachments to spread viruses or steal private details.

Scam Red Flags Quick-Glance Box

- No prescription required for medicine
- Requests for payment by gift card or wire transfer
- Promises of secret cures or quick fixes
- Urgent language: "limited supply," "act now"
- Spelling mistakes and odd web addresses
- Requests for personal information in emails
- Unfamiliar sender with generic greetings

If anything looks off, trust your gut and don't engage. Use reporting tools built into your email or browser, and visit only official health sites when searching for new treatments or advice. This way, you can enjoy the benefits of the digital world without falling into traps set by those looking to take advantage of your trust.

PRIVACY IN HEALTH APPS—WHAT DATA IS COLLECTED AND HOW TO CONTROL IT

The moment you start using a health or wellness app, you might wonder, "What exactly is this thing collecting about me?" You're not alone—this question pops up for many, and with good reason. Health apps gather a surprising amount of personal data, all in the name of convenience. Think about it: when you enter your medication list, the app remembers every drug name, how much you take, and when you take it. Sleep trackers and activity monitors log your step counts, bedtimes, wake-up hours, and sometimes even how restless you were at 2 a.m. Some apps want to know where you walk, storing location data to map your routes or count outdoor steps. Others track your heart rate, blood pressure readings, or even mood

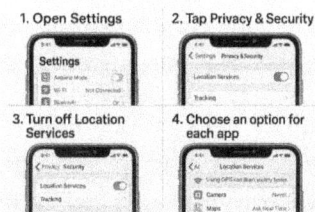

entries if you use journaling features. All this information can be helpful, but it's crucial to understand where it goes and who can access it.

Taking a closer look at privacy settings is a smart move. Each health app or device comes with its own options, hidden away in menus that can seem confusing at first. Start by opening the app's main menu and looking for "Settings" or "Privacy." Here, you'll often see a list of permissions and toggles. For example, Fitbit lets you decide if location tracking stays on; slide this setting off if you don't want your walks mapped. If you use Apple Health or Google Fit, you can select which specific health details are shared with other apps or devices—maybe you're fine sharing step counts but not your sleep history. Go through each item one at a time and ask yourself, "Do I need to share this?" If the answer is no, switch it off. Don't rush—spend a few extra minutes now so your private health info stays just that: private.

Permission requests often pop up the first time you install or open a new app. The screen might say "Allow access to contacts?" or "Enable camera?" These prompts can be confusing, especially when you're in a hurry to get started. Take a breath and read carefully. If you're setting up a medication tracker, there's usually no reason for it to see your photos or access your address book. Sometimes an app will need the camera to scan pill bottles, but it should ask only once—after that, turn the camera permission off until next time. If an app asks for something that seems odd—like calendar access for a pedometer app—pause and double-check its purpose in the settings or on the developer's website.

Reviewing your permissions regularly is one of the best habits you can build. It's easy to forget what you've allowed over time, especially when trying out new features or apps. Set aside a little time once a month—maybe on the first Sunday—to check all your health app permissions. Delete any apps you no longer use; even if they're just sitting on your phone, they might still collect data in the background. Update your passwords every few months as well, using combinations that are unique and hard to guess. If an app offers two-factor authentication (a code sent to your phone or email), turn it on for extra peace of mind.

To keep things clear and actionable, here's a simple checklist you can print or jot down for easy reference:

Printable Privacy Checklist:

- Review app permissions monthly
- Delete unused health and wellness apps
- Turn off location tracking unless needed
- Only allow access to the camera/microphone when prompted for specific tasks
- Check which health data is shared with other apps; limit sharing to essentials
- Update passwords regularly (every 3–6 months)
- Enable two-factor authentication when available

Visual cues help too—think of green icons as "safe" settings (like step count sharing with your doctor) and red icons as "risky" (like constant access to your microphone). Most phones let you see which apps have these permissions in one spot under "Settings"> "Privacy." Tap on each app for details and toggle off anything that doesn't make sense.

The bottom line is this: You have more control than you think. Don't be afraid to say no to permission requests that seem unnecessary or invasive. Your health information belongs to you, not just the companies behind these apps. If something doesn't feel right, trust your gut and adjust the settings or remove the app entirely.

Wrapping up this chapter, remember that technology should support you, not expose you. Keeping a watchful eye on privacy settings helps protect both your health and peace of mind as you explore everything AI has to offer. Up next, we'll look at how AI can keep your mind sharp and help you stay engaged with lifelong learning and connection.

CHAPTER 6
HANDS-ON WITH AI– TRY-IT-YOURSELF PROJECTS AND CONFIDENCE BUILDERS

"AI IN A DAY"–SET UP YOUR FIRST SMART ASSISTANT FROM SCRATCH

REMEMBER the curiosity and slight nerves that come with trying something new, like unboxing a kitchen gadget or a radio? Setting up a smart assistant such as an Amazon Echo or Google Nest Mini is often much easier than expected, more like plugging in a lamp than assembling a tricky appliance. With patient, step-by-step guidance, you'll see that success is just a series of small actions away. Even if you've never set up a device before, this is entirely doable.

Start by unboxing your smart assistant. An Amazon Echo is about the size of a coffee mug, while a Google Nest Mini is smaller, resembling a thick coaster. Inside, you'll find the device, a power cord, and sometimes a small instruction booklet. If the written instructions seem complicated, don't worry—we'll keep it simple. Examine your device: the Echo typically has volume buttons (plus and minus), an action button (small dot), and a button to mute the microphone (a line through a circle). The Nest Mini offers touch controls for volume and a physical switch for muting. The power port sits along the edge or the base. Plug in the device where you'd like to use it—many people prefer convenient spots like the kitchen or near a reading chair.

Once plugged in, your device will light up—Echo with a blue ring, Nest Mini with four dots—showing it's turning on and ready for setup. Wait for the lights to finish flashing. If they keep pulsing for more than a minute or so, unplug and try again. These devices are not delicate; they're made for daily use.

Next, connect your assistant to your Wi-Fi using your phone or tablet. Open the app store and search for the "Amazon Alexa" (for Echo) or "Google Home" (for Nest Mini) app—both are free. Download and open the right app. If you'd like, select large text or high-contrast display for easier reading. These apps provide strong accessibility features, like voice support and screen readers (see AARP guide).

To create your account, enter your name and email, and choose a memorable password (avoid something very simple). The app will help you find your device (which may be listed with a model number found on its bottom) and prompt you to select it. Next, enter your Wi-Fi password to connect. If you make a typo, just try again—confirmation appears on both your device (solid light) and the app ("You're connected!").

Now, test out your new assistant. Try cheerful requests: "Alexa, tell me a joke," or "Hey Google, what's the weather tomorrow?" Your assistant should respond clearly. If there's no response, check if the microphone is muted (look for a red light or mute switch). Try other basic commands: "Play music," "What time is it?" or "Set a timer for five minutes." These assistants are designed for easy, natural commands.

Explore the functions of each button using large-print photos in this book or the interactive diagrams in the app's Help section. Don't worry about experimenting—the worst case is muting the mic or changing the volume, both of which are simple to fix.

Printable "Quick Start" Checklist

- Device plugged in
- Lights steady/ready
- App downloaded (Alexa/Google Home)
- Accessibility options chosen (large text/high contrast)
- Account created
- Device connected to Wi-Fi
- First command tested ("Tell me a joke")
- Heard the first response
- Adjusted volume/mic as needed
- Location chosen for everyday use

Print and keep this checklist handy. Checking off each step demonstrates your progress. Share your setup success with friends or family —take a photo to celebrate.

With your first smart assistant running, notice how straightforward technology can be when tackled step by step. Every small action builds your confidence for next time.

CREATING A CUSTOM ROUTINE—AUTOMATING DAILY TASKS WITH AI

If you've ever wished your mornings or evenings could run a bit smoother, routines—sometimes called automations—can become a quiet ally. Routines are like little scripts you set up so your smart assistant takes care of several tasks at once. Think of them as making a pot of coffee with the press of a single button, except now, that button can also give you the weather, read your calendar, play your favorite music, or remind you to water the plants. With routines, you can bundle up the things you do every day and let technology lend a hand. For example, say you want to start each morning feeling prepared. You could set up a "Good Morning" routine that gives you the weather report, shares news headlines, and reminds you that you have appointments. All it takes is one short phrase—maybe just saying "Start my day"—and your assistant begins the sequence.

Getting started isn't complicated. In the Alexa app or Google Home app, you'll find a section called "Routines." Many people begin by selecting a pre-made routine. These templates cover everyday needs like waking up, going to bed, or leaving the house. Let's say you're interested in winding down at night. You might choose the "Bedtime" routine template. Once selected, you can decide what actions you want included. Maybe you'd like the lights to dim or switch off, relaxing music to play for fifteen minutes, and a gentle voice reminder to charge your phone before sleep. Each action is just a tap away—you add them in order, and the app shows you what's possible. Want to add a message that says "Goodnight" or a reminder to lock your doors? Simply select those actions and place them where they fit best in your routine.

Personalization is part of the fun. You can adjust each routine for your lifestyle and needs. If you prefer to wake up slowly on weekends, set your "Good Morning" routine to start later on Saturdays and Sundays. For music lovers, start your day with a favorite playlist after your morning weather update. You might even add a daily joke or inspirational quote to begin with a smile. The process is less about programming and more about choosing options from a friendly menu. Most apps use plain language and give helpful tips along the way.

Next comes scheduling or triggering your routine. You have choices: let it run automatically at a set time each day or start it manually with a spoken phrase. For example, program your "Good Morning" routine to activate every weekday at 7 a.m., so you wake up to soft lights and friendly reminders without lifting a finger.

Alternatively, use a custom command like "Alexa, goodnight" whenever you're ready to wind down—the assistant will respond by running your chosen actions in order. This flexibility lets routines fit seamlessly into your life, whether you're an early riser or prefer leisurely mornings.

Sometimes routines need a little fine-tuning, especially when trying new automations for the first time. If your lights don't turn off as expected or music doesn't play, check that all devices are connected to Wi-Fi and powered on. The app may display minor warnings if there's a problem—look for red icons or error messages. From there, tap into the settings for that routine and review each step. Sometimes permissions need updating; other times, simply restarting your smart speaker solves the issue. If a routine feels unnecessary or doesn't work for you, it's easy to edit or delete it with just a few taps in the app.

Experimenting with routines is not only practical but also enjoyable once you begin to see how much time they save. Many people start small—maybe just turning off all lights at night—then expand as confidence grows. Try adding more actions or linking routines to specific days or times as you get comfortable. Don't hesitate to experiment; nothing is permanent, and mistakes can be quickly fixed by editing or removing steps.

If something feels confusing or if you get stuck, don't hesitate to ask for help from someone in your family or a tech-savvy friend. Many community centers also offer tech support hours where someone can guide you through new features (see Wired's guide for support options). Practice makes these routines feel natural over time. Even if you're unsure at first, trying out these automations can make daily life feel more manageable—and sometimes even a bit magical—when your home responds just as you like with only a word or two.

When routines are tailored for your day-to-day life, they quietly remove small annoyances and give back peace of mind. Maybe it's dimming lights for movie night, announcing reminders for medications at lunch, or starting the coffee pot before your feet hit the floor. Each successful automation is proof that technology can adapt to you —not the other way around—making life just a little easier, one routine at a time.

USING AI TO ORGANIZE YOUR INBOX—UNSUBSCRIBE, FILTER, AND PRIORITIZE

Sorting through a crowded email inbox can feel like searching for a needle in a haystack. Unread messages pile up fast—some from friends, others from old newsletters or sales you never signed up for. It's easy to feel buried. But these days, most email programs use artificial intelligence to make inbox management less of a headache. With the right tools and a little know-how, you can clear the clutter, spot what matters, and keep your sanity.

If you use Outlook, there's a handy feature called "Focused Inbox." This clever tool pays attention to your habits and moves important mail—like notes from family or your doctor—into a tab marked "Focused." Junk mail, advertisements, and less urgent messages go into "Other." You'll find this option at the top of your inbox. Just click to switch between tabs. If you spot something in the wrong place, drag it over; the AI learns quickly and improves its sorting the more you interact with it. Gmail users have something similar called "Priority Inbox." Here, messages are automatically divided into groups: "Important and unread," "Starred," and "Everything else." Google's system looks for patterns in who you reply to, what you open right away, and which messages linger unread. Gmail marks important emails with a yellow arrow. If you see the system make a mistake, fix it with a click—future messages from that sender will land in the correct section.

Unwanted emails are almost as common as junk mail in the mailbox. You can fight back with a few simple moves. When you spot a message from a store or website you no longer care about, scroll to the bottom of that email. Look for the word "Unsubscribe"—usually in small print. Click it and follow any quick instructions.

Most reputable companies stop sending messages within a day or two. If you see emails from someone who just won't stop, use your email program's "block sender" feature. In Gmail or Yahoo, open the offending email, click the three dots or "More" menu at the top, and choose "Block." That sender's future emails will be banished straight to spam. This takes just seconds and spares you future frustration.

Create new folders or labels in email

Organizing what's left is where AI-powered labels and filters shine. Maybe you want all emails from your daughter or grandkids marked as "Family." In Gmail, click on an email from them, choose the label icon (it looks like a tag), and type "Family." Set up a filter so new emails from those addresses always land there automatically. For bills or statements, create another label called "Bills." In Gmail's settings, you can create a rule that skips the primary inbox for emails from your utility company or bank and sends them directly to your Bills folder. Outlook and Yahoo offer similar features using their "Rules" or "Filters" settings. Once set up, these folders sort themselves —AI keeps watch, so you don't have to.

Tidying up is satisfying—and keeps your inbox manageable. Set aside time every week or two for some digital housekeeping. Go through your inbox and delete old spam or unwanted promotions that slipped through the cracks. Archive messages you might want later, but don't need to clog up your view. If your inbox looks overwhelming right now, start small with just ten or twenty messages at a time.

Inbox Spring Cleaning Checklist

- Unsubscribe from five email lists you no longer read.
- Create two new folders or labels (like "Family" and "Bills").
- Move recent family emails into the correct folder.
- Set up a filter for bills or statements.
- Block at least one persistent spam sender.
- Delete all spam and junk mail older than one month.

- Archive important but inactive messages.
- Review the Focused or Priority Inbox tabs for accuracy.
- Check your trash and spam folders and empty them out.
- Smile at your tidy inbox and enjoy less stress.

You can print this checklist in large font if that helps, or keep it as a note on your phone or computer. Crossing off each task is surprisingly rewarding.

One last tip: try to keep up with this routine regularly, like watering plants or checking smoke detectors. The AI behind your email gets smarter each time you label a message or block unwanted senders. Before long, you'll notice that your inbox feels lighter and more welcoming. Messages from friends and family are easier to spot, newsletters only come if you want them, and those pesky sales pitches won't sneak through as often.

AI doesn't just sweep away junk—it helps highlight what's truly important in your digital life. With these simple steps, technology works for you instead of against you, and email becomes something you can manage with ease rather than dread. All it takes is a little practice, some curiosity, and a willingness to let technology lend a hand when things get messy.

TRYING A CHATBOT—GETTING HELP FROM AI CUSTOMER SUPPORT

If you've visited a bank, store, or hospital website and noticed a chat box pop up, you've met a chatbot. Powered by artificial intelligence, chatbots are designed to answer questions and solve problems quickly, often much faster than calling customer support. You'll see them labeled "Need Help?" or "Chat with Us," ready to assist you wherever you manage your bills, book appointments, shop online, or refill prescriptions. Instead of searching for a phone number or navigating complicated menus, chatbots provide real-time answers—anytime, day or night.

A chatbot is a virtual assistant that understands your typed questions and responds with relevant information, suggestions, or step-by-step help. Unlike websites buried under menus and links, chatbots "converse" directly with you. For example, on your bank's website, you may see a chat icon asking, "How can I help you today?" Clicking it opens a chat window—no login required initially. You're greeted by a friendly, sometimes named chatbot ("Hi, I'm Sam the Chatbot!"). Just type your question as you would to a person, such as "What is my account balance?" or "How do I pay my bill?" There are often suggested topics you can click on, covering common requests like reporting a lost card or finding branch hours.

One of the main advantages of chatbots is their tireless availability. They don't take breaks, get annoyed, or mind repeating themselves. You can use them late at night or early in the morning, and they respond within seconds. When a chatbot understands your question, it replies with clear steps or direct links. Sometimes, it offers buttons for more details or to narrow down your needs. For example, if you want to see a past utility bill, it might ask, "Which account?" or "What month?"—guiding you efficiently.

Of course, chatbots aren't perfect. If your question is unusual or needs personal attention—such as disputing a charge or rescheduling an appointment—most chatbots have clear options for connecting you to a human. Look for labels like "Speak to a representative," "Request human agent," or "Call support." Clicking one quickly transfers your chat to a live person, either by phone or in the same chat window. This handoff is smooth; you won't lose your place or need to repeat everything. Despite common concerns, using a chatbot doesn't block you from reaching a real person—in reality, it often makes that transition easier.

The best way to get comfortable is to try it out. Pick a website you already use, such as your pharmacy, if you manage prescriptions online. Look for the chat icon, usually in the lower right corner. When you open it, there might be suggested topics like "Check prescription status" or "Refill medication." If you want to know if your prescription is ready, type: "Is my refill ready?" The chatbot will respond or guide you through secure login steps and ask for any details it needs (e.g.,

"Which medication are you checking on?"). Some even let you schedule pick-up times or set reminders.

Consider another scenario: you want to track something you ordered online. Go to the retailer's website and open the chat. Type, "Track my order." The chatbot typically asks for your order number (from your email receipt) and provides tracking details, sometimes with a map or estimated delivery date. If there's a problem, it can connect you directly to customer care.

Using chatbots saves substantial time. There's no need to wait on hold or for slow email responses. You can ask follow-up questions without worrying about repetition—chatbots are patient and don't rush you. They're designed for all ages, especially helpful if technology sometimes feels overwhelming.

For those who find typing uncomfortable, many chat windows offer voice input—look for a microphone icon and speak your question. If reading small text is hard, browsers and devices usually allow you to increase text size.

If you're ever unsure whether you're speaking to a bot or a person, simply ask: "Are you a real person?" Chatbots will answer and show you how to connect with a human if needed. Use chatbots for quick, simple questions, but switch to human support for anything sensitive or complex.

Trying chatbots is a good way to grow confidence with AI and digital tools. Each successful interaction means less frustration in the future and more convenient handling of things on your own time.

EXPLORING AI GAMES AND BRAIN TEASERS FOR MENTAL FITNESS

Keeping your mind sharp and entertained is just as important as taking a walk or chatting with friends. These days, you can do a bit of brain exercise right from the comfort of your favorite chair, using your phone or tablet. AI-powered brain games and puzzle apps have become excellent companions for seniors who want to keep their memory, focus, and problem-solving skills in top shape. You've likely heard names like Lumosity, Elevate, Peak, or Cognifit. These aren't just for the young or tech wizards—they're designed to meet you where you are, with activities that feel like play rather than homework.

Open the App Store or Google Play.
Search for "Lumosity," "Elevate," "Peak," etc.

Install the official app. Create an account and allow notifications.

Start with a short daily workout. Choose a difficulty level.

Track your progress over time. Consider subscribing.

Getting started is refreshingly straightforward. Go to your device's app store—Apple's App Store if you use an iPhone or iPad, Google Play if you're on Android—and type in the name of the game you're interested in. Always check that the app's publisher matches what's expected (Lumosity, Elevate, Peak, or Cognifit should appear as the developer). Downloading from the official store helps keep your device safe from viruses and unwanted ads. Many apps ask for permission to access things like notifications or contacts. It's okay to say "no" unless you want reminders or plan to share scores with friends. For peace of mind, look for the privacy policy inside the app's settings, which explains how your information is used and helps you adjust what gets shared.

Before you play, take a moment to peek at the settings menu. Most brain-training games offer ways to make things easier on your eyes and wallet. Turn on "kids mode" or disable in-app purchases if you want to avoid surprise charges, especially if grandchildren might use your device. Some apps have a "large print" option or color schemes that boost contrast, making buttons easier to spot and read. If you ever get a pop-up asking for payment, look for a small "x" or "maybe later" button in the corner; most allow free play without spending a dime.

Once inside, the real magic begins. These games use artificial intelligence to gauge how you're doing and gently adjust challenges so you don't get overwhelmed or bored. Maybe you start with a simple word scramble or a memory card game. If you breeze through, the next round gets just a touch harder—maybe more letters or faster timers. If you hit a tricky patch and miss a few answers, the game responds by easing up, giving you a chance to succeed and feel good about moving forward. This adaptive difficulty means no one gets stuck at a level that feels impossible, and there's always something fresh to keep your interest piqued.

You'll notice a variety of activities tailored to exercise different parts of the brain. Memory games might ask you to match pairs of shapes, while attention games could have you spot subtle differences between pictures. Problem-solving puzzles sometimes look like fun little mazes or require moving colored blocks into place. Each session is short—often just a few minutes—so it never feels like a chore. Plus, the games frequently come with cheerful sounds and bright visuals for added enjoyment.

Staying motivated is easier when you see progress, and these apps are experts at encouragement. After each round, many display a graph showing how your skills improve over time—maybe your memory score is climbing, or you're reacting faster than last week. Some apps award digital badges for milestones: "Played three days in a row," "Solved ten puzzles," or "Beat your personal best." These little celebrations help keep spirits high and make it easy to track improvement. Feel free to share these achievements with family as conversation starters or just as a way to pat yourself on the back.

To get the most from these tools, I recommend setting up a simple weekly routine. Playing for about ten minutes per session, three times a week, is enough for most people. Choose mornings if your mind feels freshest then, or treat yourself after lunch with a puzzle break instead of another cup of coffee. If you forget one day, there's no penalty—just pick up where you left off. Some folks like to keep a paper calendar and mark off brain-game days as an extra reward.

Don't worry if technology feels intimidating at first. Start with one app —perhaps Lumosity or Peak—and give yourself permission to explore without pressure. There's always an option to replay easier levels if something feels too challenging, and none of these programs judge mistakes harshly. Many encourage learning from them. If you get stuck in any menu or see an unfamiliar prompt, look for a "Help" or "FAQ" section—these usually offer plain-language advice and sometimes even video instructions.

Most importantly, enjoy these moments as time for yourself—a way to stretch your thinking muscles while having fun. Over time, you may find your memory getting sharper when recalling names or shopping lists, or maybe conversations flow more easily as your attention improves. Even better, these games can be played solo or together with friends and family, making them a wonderful addition to social time or quiet afternoons alike.

PRINTING AND USING LARGE-PRINT CHECKLISTS FOR EVERYDAY AI TASKS

Staying organized makes technology less overwhelming, and a good checklist can be a powerful ally. Many people find that seeing a clear, step-by-step list right in front of them calms that nagging uncertainty, especially when learning something new. If you're someone who likes having written reminders, large-print checklists are a practical solution. They help break big tasks into smaller, manageable pieces so you never feel lost. Whether you want to remember how to set reminders, keep your inbox tidy, or stay safe online, having those instructions in large, easy-to-read print gives you confidence and control. You can tape these lists near your devices for quick reference, slip them into a notebook, or even hand them to a friend who's learning with you.

Here are a few samples you might find helpful. For setting up reminders: "Open your voice assistant app. Tap 'Reminders' or say 'Add a reminder.' Speak or type the reminder message. Set the date and time. Confirm or save. Check your reminders list." For cleaning your inbox: "Open your email app. Delete spam and junk mail. Unsubscribe from unwanted newsletters. Move important messages to folders. Review your inbox once a week." For staying safe online: "Use strong passwords. Don't click suspicious links. Check sender details before opening emails. Update your apps regularly."

Everyone's needs are a little different, so making your own checklists is not only useful—it's easy with modern note-taking and word processing apps. If typing is uncomfortable or slow, most smartphones and tablets let you dictate instead. For example, on Google Keep, tap the microphone icon and simply speak your steps; the app turns your words into text automatically. In Apple Notes, open a new note and tap the microphone on your keyboard—again, just say what you want to remember. Want to add more detail? You can type or dictate extra notes anytime. Saving your checklist in one of these apps means you'll always have it handy on your phone or tablet.

Printing your checklists is straightforward, whether you're using a computer or a mobile device. In Word or Google Docs, after typing out your checklist, look for the "Print" option—usually found under the File menu. Select "Large" or "Extra Large" font size before printing to make it easier on the eyes. If you're working in Notes or Keep, choose the share icon (a box with an arrow or three dots), then select "Print." If you don't have a printer at home, you can save the checklist as a PDF and email it to yourself or directly to a local print shop; they'll print it for you using bigger fonts if you ask. Some community libraries also offer printing services, and staff are usually happy to help if you're unsure how to send a file.

Creative uses make these checklists even more valuable. Place your "voice command cheat sheet" beside your smart speaker so you don't have to memorize phrases when trying something new. A list of common email steps can stay near your computer for reference during weekly inbox cleaning. Share your favorite checklists with friends at a senior center or church group—sometimes the best way to learn is together. Bring printed lists to tech support appointments so helpers know exactly where you're getting stuck or what you want to accomplish. You can even laminate frequently used checklists at an office supply store for durability.

Customizing lists is part of the fun and helps you adapt as your skills grow. Maybe you want to keep track of which smart home routines work best for you or record troubleshooting steps that solved a tricky problem. With each update, you're building a personal toolkit—one that's as unique as your daily routine.

If you'd like an interactive twist, jot down little notes next to each item on your printed checklist: "Tried this—worked great!" or "Ask Mary about this step." Over time, these become reminders of how far you've come and where there's room to try something different.

Chapter Wrap-Up

Checklists might seem simple, but they're powerful tools when exploring new technologies. They organize, reassure, and guide you—step by step—toward greater independence with AI and digital devices. These lists keep things clear, calm, and within reach so that learning stays enjoyable, not overwhelming. Up next, we'll explore how sharing knowledge and joining supportive communities can help tech skills grow even stronger.

CHAPTER 7
SAFETY, SECURITY, AND ETHICS—PROTECTING YOURSELF AND YOUR DATA

UNDERSTANDING PRIVACY SETTINGS—STEP-BY-STEP FOR EVERY DEVICE

EVER WORRY who's peeking through your digital windows? Just like locking your front door, privacy settings protect your digital life. They put you in control, deciding who has access and how much they see across your phone, tablet, or computer. Many skip privacy menus because they sound technical or confusing, but understanding just a few key switches greatly improves your security and peace of mind.

Privacy settings matter—they are your first defense against snooping apps, unwanted tracking, and invasive marketing. Tweak them to keep private things (like location, photos, or contacts) safe, and you'll only share what you choose, leaving settings open risks exposing personal details unintentionally. It's not about hiding; it's about making informed choices. Each device, from iPhones to Windows computers, gives you simple controls, as long as you know where to look.

Let's start with Apple. On iPhone or iPad, tap "Settings," then "Privacy & Security." Here, under sections like Location Services, Contacts, Photos, Microphone, and Camera, you choose—app by app—who can access what. For example, to stop Facebook from tracking your location, tap "Location Services," find Facebook, and set it to "Never" or

"While Using the App." To use Zoom's microphone only for meetings, go to "Microphone" and turn it on just for Zoom. You're free to change these at any time. Apple also offers an "App Privacy Report," which allows you to see how often apps check your data (see Myla Training, n.d., APA list #15).

Android devices have similar options. Open "Settings" (the gear icon, found by swiping down), then tap "Privacy" or "Permissions Manager." This menu breaks down permissions for Location, Camera, Microphone, Contacts, etc. Tap each to see which apps have access. Turn off location for games or apps that don't need it. Choose if an app can always see your location, only while in use, or never. For photos and contacts, only allow what's essential. Android also allows managing permissions by app groups, making changes faster.

Windows 10 and 11 include a privacy dashboard. Click Start, go to "Settings," then "Privacy." You'll find controls for Location, Microphone, Camera, Contacts, Calendar, Email, and more. Decide what each app can access—let a video call app use your webcam, but block others. Windows summarizes recent app activity for sensitive info, making it easier to stay updated.

It's essential to review which apps can access your info routinely. Many users request permission to simplify their setup, not because they need it frequently. Social media apps, for instance, often want location access for ad targeting, which you can deny. The same concept applies to cameras and photographs—grant access only when necessary, like for video calls or sharing pictures. Always feel free to say no or revoke permissions later.

After system updates—or every month—do a privacy checkup. Updates may reset or add new permissions without warning. Regular reviews prevent accidental exposure. Don't assume settings stay unchanged after an update.

Large-Print Privacy Checkup Checklist

Print this and keep it handy. Review monthly or after updates to keep your devices secure:

- Check each app's permissions in "Settings > Privacy" (Apple/Android/Windows).
- Switch off location access for any non-essential apps.
- Restrict the camera and microphone to only trusted apps.
- Limit contact and photo access to only what's necessary; remove from unused apps.
- Review permissions following software updates—watch for anything new.
- Look at your device's privacy dashboard or report for unusual activity.
- Remove old, unused apps—they don't need your data anymore.
- After installing new apps, check which permissions they request.
- Set a calendar reminder for your next review.

With these routines, you control your devices' features without risking privacy. No need to be a tech expert; repeat these steps to keep your digital world private and safe, sharing only with those you trust.

TWO-FACTOR AUTHENTICATION—SIMPLE WAYS TO STAY SECURE

Passwords are like house keys—essential but sometimes not enough to keep out intruders. Imagine adding a keypad to your door, so even if someone has your key, they still need a code. That's the principle of two-factor authentication (2FA): an extra layer that makes it much harder for anyone else to access your accounts, even if they know your password. Setting up 2FA is straightforward and only requires following a few steps—think of it as adding a deadbolt to your digital life.

With 2FA, logging in means entering your password plus something extra—a code sent to your phone, a phone call, or a code from an app.

This means that even if your password is compromised, a hacker can't get in without access to your second factor. It's a simple, highly effective way to increase security for things like online banking, email, or health apps.

Setting up 2FA is easy on most services. For Gmail, sign in, click your initial or photo at the top right, choose "Manage your Google Account," find "Security," and then click "2-Step Verification." Follow the instructions and have your phone nearby; Google will send you a code by text or call. Outlook and Yahoo have similar options under "Security Settings." Most online banking apps prompt you to enable 2FA in "Settings," "Security," or "Login Options," and will guide you to enter your mobile number or set up an authenticator app.

Health portals like MyChart or Apple ID accounts also offer 2FA. For MyChart, log in, go to "Account Settings," and select "Two-Factor Authentication"—you can add a phone number or use an app. Apple ID users can turn on 2FA under "Password & Security" in their device Settings. If you have trouble or concerns, most services have built-in help or support contacts.

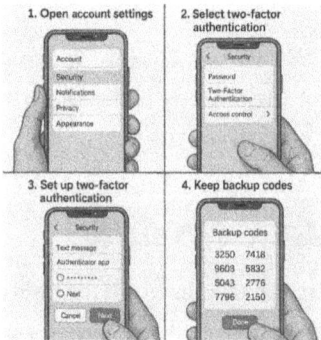

2FA methods vary. The most common is a code sent by text (SMS)—after you enter your password, you'll receive a six-digit code on your phone. Some sites can call you and read the code aloud. Authentication apps like Google Authenticator or Microsoft Authenticator generate codes right on your phone, even without cell service—handy for travelers or those with unreliable reception.

Each method has strengths and limitations. Text messages are simple but can (rarely) be intercepted if your phone number is stolen. Authenticator apps are more secure and don't require cell service, but you must keep your phone ready and familiar with the app's usage. For many, text messages are perfectly sufficient.

You might worry about losing access—what if your phone is lost or the code forgotten? Most services let you set up backups. You can add another phone number (like a landline or a trusted family member). Some sites also allow you to print backup codes to store somewhere safe, alongside important documents. These are like spare keys you can use if you're ever locked out.

If your phone is lost or broken, use backup codes if you have them. If not, contact the company's support using their website or customer service number—they'll help you verify your identity and recover access. Never share your 2FA code with anyone who asks unexpectedly; legitimate companies won't request your code unless you've initiated the call.

Make it a habit to review your 2FA settings every few months. Update backup numbers, and keep those printed codes in a safe but accessible place. Once you set up 2FA on important accounts—like email, banking, and health—you'll have one of the best defenses against online threats. You don't need to be tech-savvy; just follow the steps patiently.

Quick Exercise: Your Security Double-Check

List your most important accounts (email, bank, health portal). Choose one and check if 2FA is on—if not, set it up by following the steps above. Note where you keep your backup codes. Simple actions like these can significantly improve your digital security.

RECOGNIZING AND AVOIDING AI-POWERED SCAMS

Scammers have grown more convincing thanks to artificial intelligence. They now craft emails and texts that seem personal, often including your name, names of relatives, or personal details found online. Messages might appear to come from your bank, using impeccable English, your full name, and familiar logos. Alternatively, you may receive an email that appears to be from your doctor's office, requesting confirmation of your insurance details via a link. Such messages often look genuine at first glance.

A significant escalation is the rise of deepfake phone calls and AI-replicated voices. Imagine receiving a call from someone who sounds exactly like your grandson, discussing family events, maybe mentioning your recent birthday, and urgently requesting money while asking you to keep the call secret. Scammers use AI to mimic voices from short clips obtained online or from social media, often calling at odd hours to catch you vulnerable. If you get such a call, pause and take a breath. These scams are designed to create panic and rush you into acting before you think.

Email scams are becoming smarter. Instead of awkward grammar or obvious mistakes, scam emails now read smoothly, referencing real details, like your bank branch or a recent online purchase. The sender's address often looks almost correct, but with minor tweaks that mask imposters. Scammers typically use subject lines like "Urgent: Confirm your account or it will be closed" or "You've won a prize!" Some even pose as charities or government agencies.

There are some telltale signs to help you detect these AI-powered scams. Be suspicious of urgent demands for money, especially when payment is requested in gift cards or wire transfers. Treat requests for secrecy as a red flag, and be wary if you're told not to discuss the situation with others. If something sounds too good to be true, like winning a contest you didn't enter, it probably isn't real. Watch for slight changes in emails or phone numbers—even if everything else feels familiar—and be alert for odd payment requests such as cryptocurrency or gift cards.

If you suspect a scam, don't reply or click any links. If you receive a suspicious call, hang up—especially if you're asked for money or personal details. Don't worry about offending anyone; real friends and family prefer you to be cautious. Save any suspicious messages or call logs as evidence (screenshots are helpful), then block the sender or number—report scam emails to the Federal Trade Commission at ReportFraud.ftc.gov. For scam calls or texts, contact your phone company for blocking options, and report financial scams to your bank's fraud department.

For example, one gentleman received a late-night call from someone who sounded exactly like his granddaughter, claiming she needed bail money after an accident. The caller referenced family details, all likely scraped from social media. He nearly wired the cash but paused and texted his daughter separately to verify. That double-check prevented a significant loss—his granddaughter was safe at home.

In another case, a woman received an email from "her bank" about suspicious activity. The message used her name and an official logo, but she noticed the sender's address was a letter off. Instead of clicking the link, she called her bank directly using the number on her card and confirmed it was a scam.

To recap, look for red flags: urgency, secrecy, unusual payment requests, and out-of-the-blue requests for personal information. Trust your instincts; if a message or call feels wrong, stop and double-check with someone you trust before responding or sending money. Scammers rely on catching you off guard. Slowing down is your best protection. If you think you've shared sensitive information, contact your bank or authorities immediately—they can help secure your accounts and advise you on next steps.

Staying alert doesn't mean being fearful; it means recognizing technology has made scams more sophisticated, but also gives you more tools to protect yourself. Share these tips with friends and loved ones— a quick conversation can keep someone else from becoming a victim.

SAFE BROWSING AND APP DOWNLOADS—THE "SAFE TO TRY" CHECKLIST

The internet brings a world of convenience, but it's a bit like a busy city— there are safe, well-lit neighborhoods and some risky back alleys best avoided. Unsafe websites and shady apps can carry hidden traps: viruses, annoying pop-ups, or even programs that try to steal your information. The good news? Web browsers like Chrome, Edge, or Safari now come with built-in "AI guards" that scan websites before you visit. If you land on a dangerous site—maybe with malware or a known scam—you'll see a bright warning screen, usually red or orange, telling you it's not safe. It might say, "Deceptive Site Ahead" or "This site may harm your device." Don't ignore these warnings; they're there to stop you from stumbling into trouble. Most browsers also flag fake login pages, trying to catch those sneaky lookalike websites that aim to trick you into typing your password.

The same goes for app stores. When you search for a new app, look for "Editor's Choice," "Verified," or "Trusted Developer" badges. These aren't just for show—they mean the app went through extra checks for safety and privacy. Apps with few reviews or a high number of negative reviews should raise a red flag. Stick to the official Google Play Store on Android and Apple's App Store on iPhone or iPad. Downloading from unofficial websites is like picking up food from someone selling sandwiches out of a backpack—just not worth the risk.

Before downloading anything new, use the "Safe to Try" checklist like a mental seatbelt. First, is the app from the official store? If you spot it on a random website, skip it. Next, check if it has many positive reviews —dozens, even hundreds, from real people. Read a few reviews, especially the lower-rated ones, for mentions of bugs or privacy concerns. Is the developer reputable? Some apps share the same names as big brands; make sure the developer matches the company you expect. For example, if you want the official AARP app, see that it's published by AARP and not some random company with a similar name.

When clicking links in emails or texts, please take a moment to check where they lead. Hover over the link (or long-press on a phone) to reveal the full website address—called a URL. Does it start with "https://"? That "s" stands for "secure," and most safe sites use it. Look for a padlock symbol next to the address bar; this means your connection is encrypted. Beware of lookalike addresses—scammers often use tricks like swapping a single letter (ama-zon.com instead of amazon.com) or adding extra words (paypal.account-secure.com instead of paypal.com). If something looks odd or unfamiliar, don't click.

Some websites pop-up windows asking you to download software or run updates as soon as you arrive. Close those windows—legitimate updates come through your device's app store or settings, not random websites. Also, never download attachments from emails unless you're sure of the sender and expect the file.

Use this printable worksheet when you find a new app or website you're unsure about. Print several copies or keep one near your computer for quick reference.

SAFE TO TRY? APP & WEBSITE EVALUATION WORKSHEET

Question: Safe. Not sure. Is it from the official app store? Does it have many positive reviews? Is the developer reputable/recognized? Is there an "Editor's Choice" or verified badge? Does the website start with https:LL. Is there a padlock symbol in the browser bar? Did a trusted source recommend it?[Are there no urgent pop-up warnings/updates? If you answer "Not Sure" to any item, pause before downloading or entering information. Ask someone you trust, check with your local library tech help desk, or search for advice on reputable sites like AARP's technology section.

Staying safe online isn't about paranoia—it's about putting up guardrails where they matter most. With these small habits and tools, browsing and trying new apps becomes less stressful and a lot safer. And if in doubt, remember: it's always okay to walk away from anything that doesn't feel right.

MANAGING PERSONAL DATA—WHAT TO SHARE AND WHAT TO HIDE

Every time you create an account or download a new app, you're greeted with boxes to fill out—name, birthdate, email, sometimes even your address or phone number. These questions pop up everywhere, and it's easy to feel like you're just expected to give out all your details. But the truth is, not every blank needs your life story. Some information is safe to share, while other details should stay private unless absolutely necessary. Most apps or websites only need your name and an email address to get started. Sometimes your phone number is required for account recovery or security, which is reasonable, but sharing your home address is rarely vital unless you're ordering something for delivery. When you see requests for things like your birthday, it's often for age confirmation or to tailor ads—they don't need the year if it's not a bank or government service. As for medical information, only health apps and official portals should ever ask for that, and even then, you can often skip fields that aren't required.

The riskiest details to share are those that could be used to steal your identity or commit fraud. Social Security numbers, complete medical records, banking information—these are in a different class altogether. You should never enter a Social Security number unless you're dealing with a real bank, tax site, or government service you trust completely. Even then, double-check the website address and consider calling the company to confirm. If an app or random website asks for your bank account or routing number, close the tab and walk away; reputable companies won't request sensitive info through pop-ups or emails. Always pause before filling in these fields and ask yourself if it makes sense for the service you're using.

You can protect yourself by practicing "data minimization." This means providing the least amount of information needed to use the service. If a field isn't marked as required (usually shown with a little star or highlighted), skip it. Many websites will let you move forward without filling in extra details. When prompted about sharing contacts, photos, or your calendar, select "No" or "Ask me later." Marketers love collecting data to send ads or sell your info, so always opt out of marketing communications if offered the choice. Look for tiny check-boxes labeled "I want to receive offers" or similar language; make sure they're unchecked before clicking "submit."

You have more control than you think over what's already out there. On social media platforms like Facebook or Instagram, you can review your profile and remove any details that aren't needed—maybe you added your hometown years ago or listed an old phone number. Go into your settings or profile, click "Edit," and remove anything that doesn'tno longer serves a purpose. Many people forget about old accounts on websites they haven't visited in years—old email addresses, shopping logins, or even dating sites from long ago. These accounts can be a liability if left unmonitored, so it's wise to clean house every so often.

Deleting unused accounts usually starts by logging in (resetting the password if necessary), finding the account or privacy section in the settings menu, and looking for an option to "delete," "close," or "deactivate" the account. Some sites make it easy; others try to hide the option at the bottom of a long menu. If you can't find it, search online for "how to delete [site name] account" for step-by-step help. For old email addresses, set up forwarding before deleting if you think you might still get important messages. After closing accounts, check your other profiles and remove any reference to now-closed emails or phone numbers.

It's helpful to review your personal data every six months. Mark your calendar for a quick privacy checkup: skim through active accounts and update any outdated info or remove unnecessary details. This habit keeps you in charge and limits what's floating around online. When signing up for something new in the future, ask yourself: Do I need to fill this in? Am I comfortable with this company holding my data? If in doubt, leave it blank or opt for a less personal alternative.

Data Minimization Cheat Sheet (Printable)

- Only fill in the required fields on sign-up forms.
- Omit unnecessary details, such as middle name, full date of birth (unless required), or home address.
- Opt out of sharing contacts, photos, and calendar access unless it's vital.
- Uncheck boxes that sign you up for marketing emails or promotional messages.
- Never share Social Security numbers or banking info except with trusted banks or government agencies.
- Review your social media and app profiles every six months and delete any unnecessary information.
- Close accounts you no longer use; search for "delete account" options in settings.
- Edit profiles to remove outdated emails and phone numbers.
- Schedule a data review twice a year as part of regular housekeeping.

With these habits in place, you'll share less than you think—and keep more of your private life truly private.

TALKING ABOUT AI ETHICS—BIAS, FAIRNESS, AND YOUR RIGHTS EXPLAINED

When you hear "ethics," you might think of fairness or right and wrong. In the world of artificial intelligence, ethics means making sure technology treats everyone fairly, explains itself, and respects privacy. You deserve to know when a computer is making decisions about you, how those decisions happen, and whether you're being treated equally —but AI doesn't always get this right. For example, an app might not sort all faces correctly in family photos, especially for people with darker skin tones or glasses. This isn't just a random error—it's bias. AI bias occurs when an automated system gives better results for some groups than others, often because its training data lacked diversity.

Bias isn't limited to photo apps. You might see odd search results, off-target ads, or recommendations that don't fit you but work for others. AI often favors what's common in its data and overlooks those who are different. If a service seems to miss you or treat you unfairly, don't hesitate to use its "feedback" or "report" option. If you notice discriminatory ads or odd recommendations, submit your feedback or contact customer support to ask how the system made its choice. Companies are listening more now as people raise these issues.

Another key part of ethical AI is transparency. Companies should be transparent about their AI capabilities and purposes. For example, if you get product recommendations, you should know whether they're based on your searches, your location, or the general preferences of people in your group. You have the right to ask questions like, "How are these choices made?" or "Is this decision automated?" Good companies will answer directly. If they can't or seem evasive, be cautious about giving them your information.

Privacy rights have grown stronger thanks to new laws. Rules like the California Consumer Privacy Act (CCPA) and Europe's General Data Protection Regulation (GDPR) make it clear that your information belongs to you, not the companies you use. These laws give you powerful rights. You can ask any company what data it holds about you (a "data access request")—usually via an online form or email ("data request" is a handy search term on company sites). You can also ask the company to correct incorrect information or delete your data if you're done with their service.

If you want, say, a copy of all the data Facebook holds about you, check your account settings for "Download Your Information." It may take a while, but you'll receive a file with photos, comments, and more. If you want your data deleted altogether, look for "Delete Account" or "Remove Data" in the settings. This can be hidden, but it's your right to take your info with you when you leave.

Talking about AI ethics is good for everyone, not just you—it helps family, friends, and neighbors who might share your concerns. When your group tries new tech, ask how personal info is handled and whether the tool is fair for everyone. Bring up questions like, "Does this tool treat people equally?" or "How does it make its decisions?" These conversations can push companies to do better and help you make smarter choices. If an app can't answer basic questions about privacy or fairness, think twice before using it.

Being aware of these issues doesn't mean you need to worry about every new app. It just means you can approach technology with curiosity and confidence. You know what questions to ask and realize your feedback can shape how AI works. Fairness, transparency, and respect for privacy are basic rights—expect them from every digital service.

As you finish this chapter on safety, security, and ethics, remember: staying informed and asking questions keeps you in control. Use this knowledge as you explore the new opportunities AI offers for staying active and engaged at any age.

CHAPTER 8
LIFELONG LEARNING AND COMMUNITY— GROWING WITH AI AT YOUR OWN PACE

BUILDING CONFIDENCE—CELEBRATING YOUR AI MILESTONES

IF YOU'VE EVER FINISHED a puzzle and felt that little rush of pride, or looked at a flower you planted and thought, "I did that," then you know the power of pausing to celebrate. Technology works the same way—every success, no matter how small, deserves a moment in the spotlight. Many people don't realize it, but the real key to building confidence with AI isn't about the most significant leap; it's about recognizing each step forward. I've seen so many people light up when they set up their first smart speaker or manage to join a family video call for the first time. The world might not notice, but you should.

Milestones in the digital world can be just as meaningful as those in other parts of life. Last week, you might have set your first voice reminder for your doctor's appointment. Or perhaps you sent yourself a calendar invite using your phone—no more sticky notes on the fridge. Some of you might have finally organized those hundreds of digital photos from your last vacation into albums, making them easy to find and share. These are achievements worth celebrating. They mark real progress and show that you're learning, adapting, and thriving on your own terms.

Tracking these victories can be surprisingly motivating. I always suggest keeping a milestone chart, whether you prefer something you can print out and check off with a pen or a digital badge that pops up on your device as you complete new skills. You can make this as simple or as fancy as you like. Downloadable "AI Achievements" trackers are a great way to keep a visual record of your progress. Stick your chart on the fridge or next to your computer. There's something special about seeing those boxes fill up over time—a tangible reminder that you're not only keeping up, but moving forward.

Interactive Element: Downloadable AI Achievements Tracker

Here's a simple exercise: print out an "AI Achievements" tracker (or draw your own with colored pens). Create boxes labeled with goals like "Completed First Video Call," "Set Up a Smart Assistant," "Organized Digital Photos," "Sent a Voice Message," or "Used an AI Health App." As you accomplish each task, check it off or add a sticker. If digital is more your style, many apps let you collect badges for completed tutorials or daily streaks—these work just as well and can be shared with friends or family by text or email.

Don't underestimate the power of displaying your achievements where you see them every day. A certificate from an online tech class or a filled-out checklist pinned next to the calendar can boost your confidence and even spark conversations when visitors drop by. I know one reader who framed her "First Video Call" badge and put it on a shelf in her living room—her grandkids noticed immediately and gave her a round of applause.

Stories from others can be encouraging, too. I remember hearing from Harold, who had always felt nervous about using his phone for anything except calls. After reading, he decided to give digital photo albums a try. With some trial and error (and a few calls to his daughter), he sorted years of pictures into themed albums—one for each grandchild, one for holidays, and one for his garden. He said seeing those albums organized brought him more joy than he expected and gave him the confidence to sign up for an online art class. Then there's Linda, who used her new knowledge to join her local library's Zoom book club. The first meeting was nerve-wracking—she worried about clicking the wrong button—but when she saw herself on camera with new friends, she realized she belonged there as much as anyone else.

Celebrating these moments isn't just about self-satisfaction—it creates motivation to keep going. Consider rewarding yourself when you hit a milestone. Host a small "tech tea" with friends (in person or virtually) where everyone shares a new skill or trick they've learned. Or, send an email update to your grandchild telling them what you accomplished —they'll be proud, and you'll feel even more connected.

You might also want to treat yourself to something special—a favorite snack, an afternoon in the garden, or even just quiet time with a book you love—as a reward after tackling a challenging tech goal. Marking these occasions in your calendar can help you look back and see just how far you've come over time.

Remember, it doesn't matter if someone else picked up these skills years ago or if they seem easy to others. Your progress is unique to you, and every skill gained is another tool in your toolkit for independence and connection. So keep those achievement charts close, share your wins with people who matter to you, and don't forget to give yourself credit for every single victory, big or small.

BECOMING A TECH MENTOR—SHARING WHAT YOU'VE LEARNED WITH FRIENDS

There's something special about showing a friend how to use a gadget and seeing their relief when it finally works. You might not have set out to become the "tech person" in your circle, but if you can send a text, book a doctor's appointment online, or set up a voice reminder, you're already ahead of many folks who feel stuck. Teaching someone else, even if it's just a friend or neighbor, does more than help them—it strengthens your skills, too. This is that old "teach one, learn twice" idea in action. When you explain how to use a new app or walk through steps out loud, you notice the little details you may have missed before. You'll probably find yourself rephrasing things, thinking creatively, or even picking up shortcuts you hadn't spotted.

And every time your friend asks a question, you get the chance to look at problems from another angle.

If you're new to sharing tech know-how, start small. Offer to help a buddy set up a reminder for their medication or walk them through downloading a health app on their phone. Begin by sitting side-by-side or using speakerphone if you're helping from afar. Take it one step at a time—don't try to do everything in one go. Sometimes opening an app together is enough for the first session. The buddy system works best when both people have a cheat sheet—a simple list of steps, written in plain language, with big print and maybe even doodles or arrows for emphasis. Write this list together as you go, so it reflects the exact process and avoids any confusion later on. If you forget a step, pause and look it up together—there's no shame in double-checking! This shared list becomes a handy reference for future questions and builds everyone's confidence.

Clubs, libraries, and faith groups are perfect places to spread what you've learned. If you're part of a book club or craft group, consider suggesting a short "tech tip" at your next meeting—perhaps share your favorite weather app or show how voice assistants can play music on request. For those who like getting together in person, offer to set up a "tech help" table at your local senior center once a month. No need for fancy presentations; bring your device and invite people to ask what-ever is on their mind. You could even create a sign-up sheet for those who want extra help or prefer learning in small groups. Sometimes these sessions grow into regular meetups where everyone brings ques-tions and leaves feeling just a bit more empowered.

Patience is your best tool when helping others with technology. It can be easy to forget how nervous some people feel about pressing the wrong button or making a mistake they can't fix. Listen closely when someone explains their problem, even if it seems simple to you. Avoid jumping in too quickly—sometimes the best thing you can do is let them try first and talk through what they see on the screen. If they get stuck, offer choices rather than direct orders: "Would you like to try this together?" or "Do you want me to walk you through it again step by step?" Praise every bit of progress, no matter how minor. If someone finally opens an app without getting lost in the menus, that deserves recognition right then and there.

When fears pop up—and they always do—acknowledge them openly. Say things like, "I used to worry about that too," or "I also got confused the first time." This simple empathy goes further than any technical explanation. If someone gets frustrated, take a break or switch gears for a while. You might be surprised how often answers come after a cup of tea or a walk around the block. Celebrate those little wins: send an encouraging text after a session, jot down what went well on your cheat sheet, or give your buddy a gold star sticker if you're feeling playful.

The more you teach, the more comfortable you'll become—not only with technology itself but with explaining things in ways that make sense to different people. It turns out that sharing what you know can be just as rewarding as learning it in the first place. Before long, someone might come to you for advice on something new, and you'll realize just how far you've come since your first steps with AI.

JOINING ONLINE AND LOCAL TECH GROUPS FOR SENIORS

Many people think of technology as a solo activity, but joining tech-focused groups can turn learning into something much more social and rewarding. There's a certain comfort that comes from being in a room—or a virtual space—where everyone understands your questions because they've had the same ones themselves. In these groups, no one rolls their eyes if you ask how to download an app or wonder what a "cloud" really is. You'll find folks who are just as curious, maybe even a little nervous, but eager to share what they know. This sense of community does more than keep loneliness at bay; it creates real opportunities for connection, laughter, and the kind of problem-solving that only happens when people put their heads together. You don't have to figure out everything on your own.

Online forums built specifically for seniors learning technology are popping up everywhere, making it easier to find your people. Senior Planet's online forums are among the most trusted places to ask questions, swap tips, and join ongoing discussions about everything from voice assistants to health apps. The AARP hosts a range of tech discussion boards where you can post questions at any hour and expect friendly, patient answers from others in your shoes. Even Facebook has thriving groups with names like "Boomers and Technology," where members post about new apps, share troubleshooting tips, or encourage each other after a tech win. These digital gathering places have moderators who keep things welcoming and safe, and you can join from your living room—no need to change out of your pajamas.

Local in-person meetups, workshops, and classes still have a certain magic, especially if you like hands-on help or want to see familiar faces. Libraries are a great starting place. Many offer free tech classes for adults, sometimes with topics chosen by participants themselves. Check your library's event calendar or ask the librarian at the front desk for upcoming sessions. Community centers often run beginner workshops, sometimes even partnering with local colleges or high school students who volunteer as tech coaches. You might also spot flyers for city-sponsored classes on digital photography, basic computer use, or using AI-powered health tools. Signing up is usually simple—a quick phone call or filling out a form online—and most classes welcome all skill levels, so you won't feel out of place if this is your first time.

The real beauty of these groups is how quickly shared experiences build trust. When someone describes how they accidentally deleted all their contacts or struggled to join a Zoom call, you realize you're not the only one who feels lost sometimes. Group problem-solving often leads to faster answers and gives everyone a boost of confidence. Helping each other troubleshoot also means you pick up extra tricks that never make it into official manuals—shortcuts, hidden features, or even just the reassurance that "everyone gets stuck sometimes." There's nothing quite like the relief of hearing another person say, "I had that exact same problem last week!"

Of course, staying safe in group settings matters too. Whether you're in an online forum or at a library workshop, it's smart to use only your first name in public spaces. This keeps your identity more private without making things awkward. Avoid sharing personal contact information like your full address, phone number, or private email when chatting in big groups—save that for trusted friends you've gotten to know over time. If someone asks for details that feel too personal, it's perfectly polite to keep boundaries in place. Good group moderators will always encourage this kind of caution.

Respectful participation makes these spaces thrive. Listen as much as you talk; celebrate when others share their victories or admit their struggles. If someone asks a question you know the answer to, offer it up gently—remember how it felt before you learned that skill yourself. And if group rules exist (like muting microphones during online meetings or waiting your turn at in-person events), following them keeps things running smoothly for everyone.

If you're feeling unsure about joining a new group, try lurking for a little while—just reading posts or watching discussions until you feel comfortable jumping in. You'll quickly spot which forums or classes match your style and interests. Before long, you might find yourself offering advice to newcomers or raising your hand to try out the newest gadget at your local tech meetup.

Tech groups—online and off—aren't just about learning gadgets. They're about friendship, mutual support, and discovering that digital skills can grow at any age. With each meeting or message board post, you build not just know-how but relationships and confidence that spill into other parts of life. The next time you hear about an upcoming workshop at the library or see a post inviting newcomers to an online forum, consider giving it a try. There's a good chance you'll walk away with more than just tech tips—you might pick up a new friend or two along the way.

FINDING RELIABLE RESOURCES—WEBSITES, VIDEOS, AND AARP TOOLS

Finding good, senior-friendly resources online can feel like searching for a needle in a haystack, but it gets easier once you know where to look and what to look for. There's a growing number of websites and video channels made for people who want to learn about technology at their own pace, without all the techie talk that makes your eyes glaze over. One of the best places to start is AARP's "Technology Help" portal. This corner of the AARP website is packed with how-to guides, checklists, and straightforward videos about everything from using voice assistants to organizing photos. You'll also find a steady stream of tips for protecting your privacy, setting up smart devices, and exploring new gadgets—always explained in plain English and aimed at making life easier, not more complicated.

YouTube, believe it or not, isn't just for cat videos or watching old music performances. Channels like "TechBoomers" break down topics like "How to Use WhatsApp" or "Facebook Basics" in bite-sized videos, often using big fonts and slow, steady instructions. "Senior Tech Club" is another favorite—run by folks who understand what it's like to feel hesitant at first. They walk you through tasks step by step, sometimes repeating the process so you can follow along without feeling rushed. Both channels cover not only the basics but also delve into questions that arise as you become more comfortable with your device.

If you prefer a more structured approach, there are several free online courses designed with seniors in mind. GCFGlobal stands out for its simple navigation and clear video lessons on everything from email safety to using apps like Google Photos. Many local libraries now offer their e-learning platforms or have partnerships with services like Udemy or Coursera, giving you access to video classes at no cost. All you often need is your library card number and an internet connection. These courses allow you to pause, rewind, and go over lessons as many times as needed until the steps feel natural.

When searching for trustworthy online resources, paying attention to specific details can save you a lot of trouble later on. Start by checking the website address—sites ending in .org or .gov tend to be more reliable, since they're usually run by non-profits or government agencies. These sites are less likely to bombard you with pop-ups or aggressive ads. Look for an "About Us" page that explains who runs the site, what their mission is, and how you can contact them if you have questions. A real address or phone number is always a good sign. Before clicking around too much, scan for reviews or testimonials from other users.

Positive feedback from people in your age group is especially reassuring. If a site looks cluttered with flashy banners or keeps pushing you to download things you don't want, close it out and move on—there are better options out there.

Saving your favorite sites and videos makes future visits much simpler. In most browsers, like Chrome or Safari, bookmarking is just a click away. If you're using Chrome, go to the website you want to save, then click the star icon at the right end of the address bar—it'll turn blue when saved. You can also press "Ctrl + D" on your keyboard (or "Command + D" on a Mac) for a quick shortcut. Name your bookmark something memorable, like "AARP Tech Tips" or "Photo How-Tos." For even more organization, create a "Tech Favorites" folder and store all your bookmarks there. On an iPad or iPhone using Safari, tap the share button (it looks like a box with an arrow) and select "Add Bookmark." This way, whenever you need guidance, your most trusted resources are just a click or tap away—no frantic searching required.

AARP offers more than just articles—they have newsletters packed with updates on new tech trends, scam alerts, and easy-to-follow guides for getting more out of your devices. Signing up for the AARP "Tech Tips" newsletter means handy advice will land in your inbox automatically. If you ever feel stuck or want one-on-one help, the AARP tech support helpline is worth having in your contacts. It's staffed by patient folks who can walk you through problems step by step over the phone (AARP, n.d.). There's no need to struggle alone or wait for the grandkids to visit; help is just a call away.

Reliable learning tools are out there—you need to know where to find them and how to keep them close at hand. Curate your digital library of trusted resources so that when questions come up (and they always will), answers are right at your fingertips. The habit of reaching for these guides not only saves time but builds confidence every time you solve a problem or learn something new on your own.

USING THE GLOSSARY—PLAIN ENGLISH FOR EVERY AI TERM

The glossary at the back of this book is more than a word list—it's your toolkit for turning complicated tech jargon into plain, everyday language. The definitions are designed to be clear, printed in large letters, and use familiar words. Whenever a confusing tech term pops up in an article, app, or conversation, you can flip to the glossary. It's meant to help you, not hinder you. For example, if you're puzzled by "cloud storage," simply check the glossary: "Cloud—a safe place on the internet where your files and photos are stored, not in the sky." Instantly, technology becomes less mysterious.

This glossary adapts as you learn. You might begin with basics like "AI —A computer program that learns to help you," but as you explore more, you'll encounter terms like "chatbot," "machine learning," or "voice command." There's no need to memorize the entries—just refer to them anytime you find something unfamiliar. You can customize the glossary too—write notes, memory tricks, or even doodles in the margins. If thinking of the "cloud" as your "online photo album" helps you remember, write that in. Some people like to mark favorite pages with stickers or colored tabs, or even add a simple drawing alongside the definition. These personal touches help words stick more than plain definitions ever could.

If words like "chatbot" or "biometric security" show up in app instructions or family conversations, don't let them stop you. Look them up, read the definition out loud to yourself or with a friend, and repeat it a couple of times. Using the word in real conversation will help cement its meaning. For example, if faced with "two-factor authentication" in device directions, check the glossary: "Two-factor authentication— Extra step for logging in where you use your password plus a code sent to your phone or email." This way, you won't have to guess anymore.

The glossary is also helpful for groups. If you're part of a senior group or club, bring extra copies to share. Passing the glossary around builds everyone's confidence and encourages group learning. If you use email, consider sharing the glossary digitally with friends who are exploring smart devices. Some groups schedule "glossary check-ins" where members discuss tricky new tech words, look them up together, or add new, simple explanations. Making the glossary a group project ensures it stays fresh and helpful.

You can use the glossary in whatever form suits you best. Print the glossary pages to keep in a notebook or tape them to your computer for easy reference. Or use a digital version for quick word searches on your phone or tablet. Bookmarks, colored tabs, or sticky notes help navigate quickly. Highlight common words or add notes about the first time you used them successfully.

Stories can also help meanings stick. If you learned "app" after your granddaughter showed you how to create a grocery list, write next to "App—A program on your phone or tablet that does something helpful," a note like, "App = my grocery list helper." If "password manager" became clear only after struggling with multiple passwords, jot down: "Password manager—a safe place that stores all my passwords so I don't have to remember each one."

The glossary is meant to evolve with you—it's not rigid like an old dictionary. Mark it up, add to it, and personalize the entries to help you remember better. Don't hesitate to ask others how they explain terms—someone else's example might be what finally helps it make sense for you.

Whenever a news story mentions "machine learning" or someone asks about "voice commands," pull out the glossary and share what you know. This boosts your confidence and helps those around you feel more at ease with technology. Over time, these once-confusing terms will become a comfortable part of your vocabulary, and you'll find yourself needing the glossary less and less—not because you use it less, but because you've truly learned the language of technology.

KEEPING UP WITH AI—TIPS FOR STAYING CURRENT WITHOUT FEELING OVERWHELMED

The pace of technology can feel relentless, constantly shifting, sometimes unpredictably. News of another app, device update, or "breakthrough" in artificial intelligence is common, and it's normal to feel overwhelmed. However, you don't need to chase every update or know about every gadget. Think of technology as a buffet: pick what fits your needs and interests.

Remember, learning technology isn't a race; set your own pace. Instead of trying to keep up with everything, aim for small, manageable goals. For example, get comfortable with one new app every couple of months, or read just one article about AI each week. This incremental approach makes learning lighter and more sustainable, and you'll be surprised how much you absorb by focusing on what matters most to you.

Choose one or two trusted sources for your tech news instead of signing up for multiple blogs and newsletters. Reliable options like Senior Planet updates or AARP's tech digest can sort out what's essential for you, letting you keep informed with just one regular email or article a week and avoid information overload.

Ignore fleeting tech fads and dramatic "scare stories." Headlines like "AI will change your life forever!" or "Delete this app now!" are often exaggerated or clickbait. If something sounds urgent or incredible, double-check with your trusted source before reacting. Most of these stories have little substance and are meant to provoke unnecessary worry or excitement.

Talking tech with someone can make staying current easier and more social. A "tech buddy"—someone also interested in learning—can offer support and encouragement. You might catch up monthly to share tips, discuss which apps made life simpler, or laugh about your mishaps. If you're in a club or community, consider organizing a regular "tech check-in" where everyone shares a new tip or question. These informal chats foster honest conversations about what works and what doesn't, helping you spot tech updates worth paying attention to.

Advice often comes best from peers just ahead of you on the tech curve rather than TV experts. Don't worry about forgetting things between meetings; that's why having someone alongside you makes a difference. Over time, you'll get better at identifying what matters.

For device updates, there's no need to install everything right away. Many updates are helpful—improving security or fixing bugs—but not all will affect your day-to-day use. If you're unsure, consult a tech buddy, your group, or the device provider's helpline. Usually, there's no rush; wait until you feel comfortable.

Set boundaries with technology for your well-being. You can manage which notifications you receive, mute overwhelming news alerts, and log off to avoid information fatigue. Take breaks from screens and skip topics that don't matter to you. Focus on tech features or apps that help with your routines, hobbies, health, or social connections.

Try keeping a notebook or using a phone's notes app to jot down tech questions as you think of them. Next time you're learning or chatting about tech, look up those answers or ask someone for help. This intentional method avoids panic and makes learning smoother.

There will always be a new tech development around the corner, but you don't need to chase them all. Steady growth at your own pace is the goal—just learn what's useful and set aside what isn't relevant right now.

In summary, staying current with AI isn't about knowing everything—it's about feeling comfortable asking questions, exploring new things when they interest you, and knowing when to say, "that's enough for today." When you set your own terms, technology becomes manageable and even enjoyable. Next, we'll share real-life stories showing how these small steps lead to meaningful change, reminding us that progress is about making daily life better, not keeping up with everything new.

KEEPING THE GAME ALIVE

Now that you've learned how AI can make daily life easier, keep you connected, and open new doors, you hold the tools to make technology work for you. The next step? Share what you've discovered so others can take the same journey.

Leaving your honest opinion on Amazon is like pointing out a shortcut on a walking path—you save someone else time, confusion, and maybe even a little frustration. You're showing other beginners where to find clear, friendly guidance without feeling overwhelmed.

Every time we pass on what we've learned, we keep the conversation about AI alive. Your review helps me keep building a space where people can learn, grow, and feel at home with technology—no matter their starting point.

Thank you for helping make AI feel less like a mystery and more like a trusted friend.

To make a difference, just scan the QR code below or visit this link:

[https://www.amazon.com/review/create-review?&asin= B0G3WVPNY8]

— *Gwen Blake*

CONCLUSION

First off, let's take a moment—yes! You've made it to the end of a book about AI and smart devices, and that's worth celebrating. Whether you read every word or hopped around to the parts that fit your life, you took a real step toward something new. I wrote this book especially for you, not for the folks who already have a drawer full of gadgets and know every setting by heart. My goal has always been to help you, as someone 60 or better, feel more confident with technology. I wanted to use plain language, real examples, and a friendly, supportive tone—no jargon, no talking down, and no expecting you to be a tech whiz overnight.

Let's look back at what you've accomplished. We started by taking the mystery out of AI. You learned that artificial intelligence isn't just the stuff of sci-fi movies—it's the quiet helper behind the scenes, suggesting reminders, grouping your photos, or organizing your inbox. We discussed how smart speakers like Alexa, Siri, and Google Assistant can feel like new friends, ready to help with a simple question or voice command. You saw that you don't need to memorize complicated steps to get real benefits from these tools.

We dug into communication—how AI makes it easier to stay in touch with family and friends through video calls, smart messaging, and safe email. We went through simple ways to organize your digital world, from sorting photos and creating albums to making sure those precious memories are easy to find and share. We explored how AI can improve your health routines—whether it's remembering medication, tracking steps with a wearable, or making telehealth visits feel less stressful. We've invested time in security and privacy, ensuring your digital life is safe and your personal information is protected.

Hands-on projects gave you the chance to try things for yourself, with checklists and step-by-step guides designed to make each task feel manageable. We talked about connecting with others through tech groups, sharing what you learn, and finding trusted resources—because you're not alone in this. There's a whole community of seniors learning and growing together.

Here's the biggest takeaway: you are more capable than you might think. AI is not some distant, complicated thing—it's already helping you, often without you realizing it. If you've ever wondered how your phone knows which photos are from your granddaughter's birthday or how your email seems to know which messages are junk, that's AI quietly working for you. Safe, independent use of technology is absolutely possible, and learning is a journey you can start (or continue) at any age.

I want to recognize your progress, no matter how small it feels. Maybe you set up a reminder for the first time. Perhaps you joined a video call with family. Maybe you just read through a checklist and thought, "I could do that." Each step counts. Every time you try something new—even if it feels awkward at first—you're building confidence and independence.

If you're still a little nervous, that's perfectly normal. Uncertainty is part of learning, not a sign that you're "not good with tech." Mistakes will happen. Devices sometimes act up. That's why I packed this book with tips, checklists, and troubleshooting ideas. Feel free to revisit them anytime. They're here to catch you if you stumble, not to judge you if you need a second (or third) try.

I also encourage you to keep exploring. Go back to the hands-on projects or checklists when you're ready. Try a new app or feature. Join a local tech class or an online group—libraries, senior centers, and resources like AARP's Tech Help are great places to start. If you enjoy learning with others, share what you know. Teaching a friend or family member isn't just generous—it actually helps you remember and enjoy what you've learned.

If you want more guidance, check out the list of trusted websites, videos, and tools I included. AARP's tech support, YouTube channels for seniors, and your local library's digital classes are all safe bets. And don't forget the glossary at the back of this book. It's there to turn confusing tech jargon into plain English whenever you need a quick refresher.

You're now part of a growing community of seniors who are saying "yes" to technology. Your curiosity and willingness to try—even if you started out feeling unsure—are powerful. You're setting an example, not just for your peers, but for younger generations who need to see that learning never stops.

So, what's next? Don't let this be the end of your journey. Pick one thing—just one—from this book. Maybe set up a new reminder using your voice assistant. Consider printing a photo and sharing it with a friend. Maybe talk to someone else about what you've learned, or sign up for a tech help session at your library. Every time you try something new, you become a bit stronger, more connected, and more independent.

Remember, you're not alone. I'm cheering you on, and so is a whole community of learners. Keep your curiosity alive, keep asking questions, and keep sharing your wins—big or small. The digital world has a place for you, and you've already proven you belong here.

Here's to your next adventure—one click, tap, or "Hey Siri" at a time.

GLOSSARY OF TERMS

Algorithm

What it means: A step-by-step set of instructions a computer follows.

Memory cue: Like a recipe in a cookbook, but for a computer.

Example: When Facebook shows you birthday reminders, it's following an algorithm.

Artificial Intelligence (AI)

What it means: Technology that lets computers "think" and learn from experience.

Memory cue: Like teaching a grandchild how to bake cookies — they get better each time.

Example: Netflix recommending shows you might like.

Automation

What it means: Tasks done by machines or computers without your help.

Memory cue: Like a coffee maker that brews at 7 a.m. every morning.

Example: Your email automatically sorting out spam.

Big Data

What it means: Huge amounts of information that computers study to find patterns.

Memory cue: Like a giant library filled with every book ever written.

Example: Health researchers studying millions of medical records to spot trends.

Chatbot

What it means: A computer program you can "chat" with online.

Memory cue: Like a helpful store clerk, but on a website.

Example: A pop-up window on your bank's website asking, "How can I help you today?"

Cloud Computing

What it means: Storing files and programs on the internet instead of your own computer.

Memory cue: Like keeping your important papers in a safe deposit box you can open anywhere.

Example: Using Google Photos to store pictures instead of saving them only on your phone.

Data

What it means: Information a computer uses — numbers, text, pictures, or videos.

Memory cue: Like ingredients you gather before cooking.

Example: Your step count from a fitness watch.

Deep Learning

What it means: An advanced kind of AI that mimics the human brain.

Memory cue: Like practicing the piano until you can play without looking.

Example: How voice assistants like Alexa get better at understanding your accent.

Facial Recognition

What it means: Technology that identifies people by their face.

Memory cue: Like recognizing your neighbor at the grocery store.

Example: Your phone unlocking when it sees your face.

Generative AI

What it means: AI that can create new things — writing, images, music, or video.

Memory cue: Like an artist who can paint in many styles after studying thousands of paintings.

Example: AI creating a greeting card design from your description.

Machine Learning (ML)

What it means: Computers learning from data without being told exactly what to do.

Memory cue: Like learning to ride a bike by practicing, not by reading instructions.

Example: Your email learning which messages are junk and which are important.

Neural Network

What it means: A computer system inspired by how the brain works.

Memory cue: Like a web of roads connecting towns — each path helps information travel.

Example: AI figuring out if a photo shows a cat or a dog.

Natural Language Processing (NLP)

What it means: AI's ability to understand and respond to human language.

Memory cue: Like talking to a bilingual friend who translates instantly.

Example: Dictating a text message and having your phone type it out.

Prompt

What it means: The words or instructions you give an AI to get a result.

Memory cue: Like telling a waiter your exact order.

Example: Typing "Write a funny birthday poem" into ChatGPT.

Speech Recognition

What it means: Technology that understands spoken words.

Memory cue: Like a secretary taking notes while you speak.

Example: Saying "Call Mary" to your phone and it dials her.

Training Data

What it means: The information AI learns from before it can work well.

Memory cue: Like practicing with flashcards before a quiz.

Example: Feeding thousands of dog photos to AI so it learns what a dog looks like.

Virtual Assistant

What it means: AI-powered helpers that follow voice commands.

Memory cue: Like having a personal secretary on call 24/7.

Example: Asking Alexa for the weather forecast.

Voice Cloning

What it means: AI technology that can copy a person's voice.

Memory cue: Like a skilled impressionist mimicking a celebrity.

Example: AI reading a message in your own voice.

Wearable Technology

What it means: Devices you wear that collect information and sometimes use AI.

Memory cue: Like a watch that's also your health coach.

Example: A Fitbit tracking your heart rate during a walk.

Wi-Fi

What it means: Wireless internet connection for computers, phones, and smart devices.

Memory cue: Like an invisible cord connecting you to the online world.

Example: Using your tablet anywhere in the house without a cable.